Emmalien's Constellation

Clair Voyant

ISBN (Paperback): 979-8-9927754-3-3
ISBN (eBook): 979-8-9927754-2-6

Chapter 1

"I don't want you to go!" Maryann protested, fighting tears. "It feels like you just returned from the sea yesterday!"

"I don't want to go either, but the job pays well," Edmond Conway tried to reassure his wife. "This was a lucky break, really. Being the only ship in the harbor when they needed transport. It was as if I was meant to take this job."

"That's not luck." She didn't hide her disappointment at all. "That's a curse, as far as I am concerned."

"Please, Maryann." Edmond lowered his voice to sound more soothing, "I do this to give you the best life that I can. A life you're worthy of."

"I don't care about money," she snapped, "I care about you!"

"I'll be fi—"

"I have nightmares every time you're away!" she told him. "Did you know that? I wake up in terror most nights, afraid you have fallen into the sea. Afraid you've gotten lost and that you'll never return to me!" He didn't respond, only stared at her with a frown. She went on, "You don't know how many days and nights I stand by the harbor just waiting to see your sails come over the horizon."

"I rush back to you," he said. "Emmalien's Constellation cuts through the waves like warm butter. I'll return to you again, just as I always do."

He could see his wife was truly upset. He hated to upset her. He had only ever wanted to make her smile. "I do it for you," he went on when she didn't say anything. He rested his callused hand on her swollen stomach. "I do it for both of you." She was seven months along now, which only added to the stress of her being alone.

She rested her hands over his, "I understand," she said, "but we need you. I need you. I don't want you to miss the birth."

"I will try my very best to make it back in time." He wrapped his arms around her and gave her auburn hair atop her head a gentle kiss. "I have to be back in time to name her."

"Her?" She smiled. "You think it is another girl?"

He felt a small smile tug at his lips. He wanted a healthy child, but he couldn't help but hope for another daughter. The pain of losing Emmalien two years ago still broke his heart. She had only been five. A sickness took her while he was away at sea. He never forgave himself for not being there when she needed him most, and for not being there to hold her as she faded.

While she was gentle and kindhearted, he knew his wife still held onto this fact, too. Ever since they lost their precious baby girl, she hated him going out to sea. Before, she seemed more understanding. She knew the life she signed up for when she married a ship captain. That understanding seemed to fade more with each passing day.

Perhaps losing their daughter showed her just how frail and fleeting life could be. She had already lost so much. To lose him would mean to lose everything she valued in life. To lose him would also mean losing the father of their newest addition to the family.

"I think it is either a girl or a boy," he guessed playfully, but he could see it in her eyes. She longed for a daughter just the same. It would never replace Emmalien—nothing could—but it would feel like their family was complete to an extent again. To have little feet running around the house again.

"You're something," she joked, but her smile faded as she lingered on a thought within herself. Finally, she made it known: "I still don't understand why you can't just take me with you."

"You know I can't," he reminded her yet again.

She mocked his voice now, "Bad luck to bring a woman aboard."

"It is." He raised his eyebrows. "Every sailor knows it. There's a reason it's been set in stone, you know."

"Silly superstition." She rolled her eyes.

"It's for the best." He rubbed her tummy again. "Trust me when I say you don't want my men delivering a baby on a rocking ship. You'll be better off with the women—the midwives."

"As you say." She pouted, but ultimately, she did know he was right. She just hated that he was leaving in the morning. This trip came with no warning; she had no time to really prepare. She was surprised the ship had already been loaded and that the cargo was freighted before she even knew the trip was to happen. It was as if the crew couldn't wait to leave while she dreaded the whole thing.

"You call my rules superstition," he joked, "yet you function through women's intuition."

"I'm usually correct," she defended herself. "Just like I am right with this. This trip is no gift; it's a curse. You'd better come home to me in one piece."

"You worry too much." He kissed her lips and added, "That is why I love you."

She rested her head on his shoulder, clinging to him and taking in his scent, knowing she wouldn't be able to do this again for a month or more. It pained her heart, but she knew there was no swaying him. All she could do was soak up the time she had left with him before he departed. And she meant to.

•••

The morning was rough. She didn't want the sun to come up, yet it did anyway, announcing the horrid departure was at hand. She reached her hand out next to her to find the sheets cold and empty. Her heart dropped into her stomach.

Had he left her without saying goodbye?

She flung herself from the bed, only for the weight of her stomach to cause an ache in her lower back. She ignored the pain and rushed from the room.

There he was. Edmond Conway sat at their table, drinking a cup of coffee while reading the paper. He looked up at her, noticing her distress. "Are you alright?" He got up to go to her.

She threw her arms around him and kissed his face all over.

"Easy," he gave her a gentle squeeze, "or I will have to take you right back to the bed, woman."

"I thought you'd left me already." She sounded so sad still.

"Not yet." He chuckled at her. "It will be an hour before I go. The finishing touches are being prepared on the ship."

"You truly have to leave me?" she asked for the hundredth time, still in disbelief.

"I'll be back before you know it. Care to see me off?"

"You know I'll watch you disappear beyond the horizon," she said sadly.

"I'll watch for you, too."

•••

He always longed for home when he was away—longing for his wife's embrace. Yet, at the same time, he longed for the sea. He longed for the crash of the waves against the wooden hull of Emmalien's Constellation, the cool salty mist that would layer over his warm face and clothes, and the steady rock of the boards beneath his feet. He almost didn't even feel them anymore. This was his second home.

The shore was shrinking in the distance, the people beyond any visibility by now. The first steps were the hardest part of the journey. Watching home grow smaller and smaller in the distance while part of you simply wished to return home to your cozy bed. Alas, there was work to be done.

The ship was busy with the crew maintaining everything. The deck buzzed with life. These were men he had known for many years: men he trusted, men that admired him—all except for the two strangers he was transporting. He didn't know them from Adam, but he figured they must be important to be willing to pay so much. Money like that could take care of his family for many years. He may not have to take another job so soon. His wife may have been upset by this job, but she would thank him later. She appreciated all he did for them, anyway.

Edmond noticed the two men staring out over the water. He came down further onto the deck to address them.

"I hope you found your quarters to be acceptable," he said to them, but he doubted it. These men came from serious money. It wasn't likely they were truly satisfied with the leaking, rat-infested room they would have to share.

"It's perfect." One of the men smiled. He had an accent that Edmond had no way of pinning. He could have been from anywhere. Despite his curiosity, he thought it would be rude to ask them where they were from.

Although Edmond was sure they were only being polite, he was happy to hear it.

"We can't thank you enough for your willingness to help us," one of the men said.

"I know how it feels to be far from home," Edmond assured them. "I am happy to help."

"When is mealtime?" the other man asked. "I have quite an appetite."

"We've only just set sail." The captain laughed, but he saw neither of them were amused. "You'll need to take it up with DeVitas. He is the cook on board. He should be able to help you with that."

The captain left them to go about their business. He checked the sun again, shining ever so brightly as they sailed to the east. Then he checked the locket around his neck that had pictures of his wife and little Emmalien in it. This was his comfort when he felt stressed or unsure. He always intended to carry them everywhere, always keeping them near his heart.

That night, he updated his logbook. Cataloging their position, course, and speed, then adding more.

June 7, 1834

Not a cloud in the sky. Expecting clear weather for the foreseeable future. The sun is bright and there is a healthy breeze.

Starting out smoothly. The two men are adjusting well. I imagine this isn't their first time traveling by ship, though not likely in a vessel such as this. They likely have had transport of a much higher class.

Hate to be torn away from Maryann at a time that is so important, but I couldn't turn down the salary. The crew was eager to take the job as well.

Let's see how well they hold up for the duration. Men that have money like them aren't typically hardened to the sea.

He turned in for the night and allowed the ship to rock him to sleep.

Chapter 2

Two weeks had passed, and the captain had already heard his fair share of complaints against their passengers.

"They are always down in cargo."

"They scream at the rats."

"They make all manner of noise in the night."

"They keep speaking in their native language—it is as if they are conspiring."

"They act too high and mighty."

This night, the sun had only just set before DeVitas was sitting across his desk in his captain's quarters.

"They eat all throughout the day, taking all the fish from our stores. There are designated meal times! I cook when I cook." DeVitas was more than irritated. "How can they eat so much and still be so slim, I wonder? They don't even cook the fish they take. They're animals to be eating it raw!"

Edmond found this to be a little troubling, but he tried to reason with the cook. "They did bring those fish with them. Rightly, those belong to them."

DeVitas sounded offended, "They told me they brought those extra fish for *everyone*. That it was meant to be shared. Tell me why they are the ones eating it all before anyone else has a portion? Do they understand how rations work?"

"There is enough for them," Edmond said calmly. "We only have a few more days in their company. Then, you will never see them again."

"You defend them?" DeVitas was shocked.

"Keep your head on straight." Edmond went on to remind him, in case he had forgotten, "Once this job is done, you can buy the whole ocean's worth of fish."

This was the route the captain usually had to take in order to calm his crew: reminding them of the payout. Reminding them that, no matter how much they didn't like these men and these arrangements, it was going to be worth it in the end.

"How do you know they are good for it?" DeVitas asked. It was the first time anyone had asked Edmond this. He wasn't expecting it.

"Because if they aren't, they'll have more than enough fish to eat when I drag them back out to taste the sea." Edmond wasn't a particularly violent man, but the idea of potentially missing the birth of his child because he was out getting swindled didn't sit well with him. These men promised riches. They had better deliver.

"I'll throw them overboard myself if that's the case," DeVitas seconded his captain's sentiment.

The captain chuckled. "Shouldn't you be cooking now? The men will riot if they don't get full bellies."

"Aye," DeVitas complied. He nodded, then made his exit.

Edmond rubbed his temples for a moment, feeling his headache slowly fading. Another crisis barely diverted. Keeping the crew in good spirits to perform their duties was paramount.

He needed some fresh air.

Edmond went above to the deck for a stroll. The boards creaked underfoot, the air was cool, and the wind passed gently through his salt and pepper hair and beard.

He went to the side of the ship, casting his eyes out to the endless waves around them. There was a certain peace to being so far out. It was a strange small feeling to know that even though the waters stretched out as far as the eyes could see, that was only the top, flat of the ocean. There were miles just under their very feet that stretched even deeper into the depths below.

He glanced up at the stars; yet another ocean above that stretched even further. Meeting the horizon, just to go beyond and wrap itself around the world.

He smiled when he thought of Emmalien again. She was so used to the ocean that she had even thought the sky was an ocean: a brilliant blue of deep waters in the heavens. That each star at night was a lighthouse in the distance.

Whenever he used to take her out at night on the ship he had named after her, she would stand up front in excitement as they sailed. She wanted so badly to see where the horizon ended. She would have made a fine sailor. Full of excitement for adventure.

"They're stunning."

Edmond heard the accent from behind him as his fond thoughts were interrupted.

He looked back to see one of the men standing behind him, looking up at the stars. "You don't see so many stars in the cities. Only out here can you really appreciate the expanse of them."

Edmond nodded in response.

"Do you have a favorite?" the man asked. "Star, I mean. Is there one you look towards every night?"

Edmond felt a tightness in his chest as he thought about whether to tell him or not. Ultimately, he didn't see the harm in sharing it.

He pointed up above them to the sky. "There. Not just one star—a constellation. Not one in any book. I don't even know if anyone else sees it as I do." He brought his hand back down now to hold onto his locket. "I see her shining smile. Her playful dimples. Her bright eyes. My Emmalien. She comes to see me every night."

The man had looked to where Edmond had pointed, but whether he saw it or not wasn't clear. "Ah." He nodded. "A woman who waits for you back home?"

Edmond felt the pain in his chest again. He frowned before he answered, "My daughter."

The man's eyes widened. "I see now. Emmalien's Constellation. That makes sense. You must love her very much."

"I always will," Conway confirmed with zero hesitation. "Do you have a family? Is that where you are going now? Back home to them?"

"Oh no," the man scoffed. "This is a business trip. My colleague and I are traveling for work."

Edmond recalled the large cargo they had brought on board. The captain was curious now. They had so much money, their job must be one of great importance. "What line of work are you in?"

"I'm a researcher," he said quickly. "My line of work is collecting evidence."

Edmond was surprised by that. He didn't know that paid so well. "A man of science."

"Yes!"

"What manner of evidence do you mean to find on this journey?"

The man seemed hesitant to answer. After a moment of consideration, he finally spoke. "For now, the contents of my research are classified, but you will know soon enough. The whole world will!" He sounded excited now. For some reason, his tone made Edmond uncomfortable, though he couldn't understand why.

His wife's words echoed in his mind. She had called this trip cursed. With the unsettling feeling twisting his gut, he could only hope that she wasn't right.

Chapter 3

June 29, 1834

The men treat this like a vacation. It is almost funny. I caught them sunbathing on the deck today. The weather is perfect for it, so why not? This will be the easiest salary I have ever earned. The entire trip is going incredibly smoothly. Had I not such pressing matters at home, I could call this a vacation trip.

Spotted some whales today. Dark gray with white spots. They came close to the ship. I wonder if they know what a ship is. Either way, the guest passengers loved the sights. They're researchers, after all. I do wonder what exactly it is they research.

We should reach our destination by morning. The course is clear.

With that entry, Captain Edmond Conway closed his logbook and prepared himself for sleep. He looked forward to the warm blankets. His job was just about complete.

Soon.

Soon he would be back home.

He did the same as he always did before bed. He closed his eyes and thought of Maryann: her bright green eyes, her freckles, her dimples, her sweet kiss, her delicate touch, and her warmth next to him in bed. He grasped the locket and smiled while he imagined her head resting on his burly chest. Her hand would rise and fall as he took in breath. He would run his fingers through her hair until she would snore and drool onto his chest.

A loud and powerful crack of thunder vibrated every wooden board of the ship.

Edmond flung himself up in bed, realizing in a half daze that he didn't even know he had fallen asleep. How long had he been out? Better yet, was that noise truly thunder, or was it something else?

Lightning flashed outside of his windows, causing the whole room to light up as brightly as day. The bolt was followed by another monstrous clap of thunder that seemed to shake his very bones.

A storm?

That was when he heard it. The hard rattle of the windows as a torrent of rain came against the glass. It sounded like heavy marbles being poured onto a mirror. He half feared the glass would shatter.

How could it be storming?

Suddenly, the ship jerked so violently that Edmond was thrown from his bed. He hit the floor hard, taking his blankets with him. The ship rocked back in the other direction hard, flinging him back against his bed. He would likely have a bruise from that later.

He heard men screaming outside his door. It wasn't likely anyone was sleeping now.

He had to check the crew. He had to make sure the ship could withstand the storm. He rushed out to meet the craze. The crew members were all running in various directions. He had to get to the deck.

He followed a couple of his men up above. The deck was being pelted with rain that hit like hail on his arms and face. He could barely open his eyes because he was being hit in the face so much.

He shielded his eyes as best as he could and saw his men at work. He had sailed with these men for so long, they moved like clockwork. They knew the routine during a storm.

A flash of lightning cracked across the sky, revealing a huge wave headed straight for them from the side. At this rate, they'd be broadsided.

He yelled the warning over the rain. The ship began to turn to face the wave head on. They had to hit it at the right angle in order to make it through.

The warning paid off. The bow came into contact with the wave first. A rush of water spilled violently over the bow and across the deck. The ship creaked and groaned as it bent and bowed along with the incoming waves, sounding as if any second, it would crack right in half.

The wind came as sharp as a knife. A loud clap was heard as one of the sails ripped from the strain. The men were practically fighting for their lives to hold on while getting the sails closed up. This storm must have taken everyone by surprise.

Edmond took his place at the helm of the ship. He had seen more than enough storms to know how to steer in them.

Another wave was approaching.

His men braced for impact.

Again, the ship groaned, and a deep roar of thunder sounded as water rushed up along the sides of the ship.

Trying to keep her steady wasn't easy. The storm seemed to be picking up instead of slowing down. It was only getting worse, and, consequently, harder to see. Edmond was soaked completely to the bone. His hands threatened to slip off the helm with all the moisture.

He heard his men yelling but couldn't make out what they were saying. He didn't have to guess for long.

The lightning showed him.

It was another wave, except this wave was unlike anything Edmond had seen before.

It was a monstrous wave that had to be at least 80 feet high.

It was less like a wave and more like an incoming wall of water that seemed to stretch all the way up to the thick clouds above.

There was no straddling this wave. This would engulf them.

"Hold onto something!" Edmond yelled as loud as he could to his crew.

The wave seemed to approach in slow motion. In fact, everything seemed to be going in slow motion. It was as if Edmond could see every individual rain drop that hit the deck around him.

It was suddenly silent. So silent, he could have heard a pin drop. All he heard now was the beating of his own heart. The world was muted around him. Silent screams escaped the mouths of his crew, but he had no way of hearing them.

How long was the world silent? Seconds? Minutes? Hours? It felt like a lifetime.

During this lifetime of silence, he saw Maryann's smiling face with the warm sunlight flowing behind her.

He saw little Emmalien running through the grass in bare feet to give him a flower she had just picked.

He saw Maryann holding her stomach tenderly while singing softly into the night to soothe the kicking.

The very second the tip of the bow made contact with the rogue wave, the world returned to normal in an instant. A cacophony of sounds hit him all at once. The pounding rain against the wood, the crack of thunder overhead, the screams of terror all around, the immense groaning of the ship, and finally, as he gripped the helm with all his might…

Finally...

The rush of water all around them swelled into his ears, followed by the sound of muffled bubbles.

He was weightless, until everything went black.

Chapter 4

Light bounced around behind Edmond's eyelids. Muffled voices could be heard faintly in and out. It was as though he had cotton in his ears. He heard the creaking of the wood again and the soft splash of the waves. Everything was finally becoming lucid.

"Captain?" he heard clearly.

He felt a weight on his eyelids. He had the cool, damp wood of the deck under his face. He could feel that he was lying down on his stomach. His body felt like a rag doll. He could feel his limbs, but he couldn't find the strength to move them on his own. He let out a long, low groan.

"Captain? Are you okay?" the voice came again.

He felt a hand on his shoulder, giving him a gentle shake.

"Maryann?" he croaked, his throat stinging with the burn of salt.

"Come back to us, Captain," the voice called to him.

After much strain, Edmond opened his eyes. They stung, too. He had to blink several times to clear the burn and his vision.

He could see again. Willas was the one shaking him and calling out to him. His dirty blonde hair fell in clumps over his forehead. He let out a short laugh when he saw Edmond coming to.

"There you go!" Willas slowly helped Edmond sit back up and leaned against the side of the ship. "You look pale as a corpse, Captain! Thought you were one for a second!"

"I feel like one," Edmond coughed. He looked around. He assumed the sun had risen, but he couldn't see it anywhere. The clouds overhead were so thick that the sunlight couldn't penetrate it. While it was bright enough to see, he couldn't point to where the sun might be coming from.

The ship rocked steadily beneath him for the first time, making him feel nauseous. "Where are we?" The captain used the side of the ship to pull himself up. His legs shook like a baby deer. He looked around at the sea. There was no land in sight. The waves were calm again, lapping at the side of the ship that he was surprised was still in one piece.

Willas looked around himself before answering, "I don't rightly know, sir."

The captain could hear more voices. Men scattered along the deck.

"How is everyone?" the captain asked in hopes of good news.

"We lost Kaufer last night." Willas frowned sadly. "I saw it. The storm swept him right off the deck. He is lost to the ocean."

That hit Edmond hard. He would have to tell Kaufer's family. They would be heartbroken. This was supposed to be a quick and easy job. The freak storm came out of nowhere.

"Piles is wounded," Willas went on. "The bone broke clean through his leg. Halcreek tended to him, but it's clear that he isn't going to be walking anywhere any time soon." He watched the captain's face darken, yet he had to continue. "We did a roll call. It seems the storm and the wave managed to take off with four of our men."

"God be good." Edmond dug his fingers so tightly into the wood that they turned bone white. "Who?"

"Kaufer, Ashburn, Mattocks, and Gally," Willas said. "I saw Kaufer go, myself. The rest… they either went overboard or they're lost on board somewhere."

Each name was a hit to Edmond's heart. Good men were lost, and there was nothing he could have done to prevent it. It wasn't fair. Their families didn't deserve to lose them. He was dreading hearing the cries of sorrow when he returned to deliver the news.

"The passengers," he suddenly remembered. "Are they safe?" It may have seemed shallow, but if those men died, the entire trip was worthless. At this point, he could still give the salary portion to the families of the lost. Without that payment, they died for absolutely nothing.

"Aye." Willas nodded. "They hid out in their quarters the whole time." He snickered. "You should have seen them when they came out. Never seen men shaking so much before."

"How is the ship? Damages?"

"The sails could use some work," Willas informed him. "We had some leaking in the hull, but repairs are already being made."

"Good." Edmond knew Willas would handle everything. It seemed he already had done most of the work while Edmond was busy napping on the deck. How he managed to not be swept away, he couldn't fathom, but he was thankful. He could already hear Maryann's voice in his mind, nagging and scolding away at him for bothering to take this job. She would be livid when he told her about this incident.

"My head is killing me." Edmond pinched the bridge of his nose.

"You took a bit of a hit." Willas pointed to the deck where the captain had been lying. There was a smear of blood on the wood. Edmond must have knocked his head somewhere during the storm. He rubbed the back of his head, feeling blood crusted in the back of his hair.

Edmond limped his way across the deck. His men, while in low spirits, were all relieved to see their captain back on his feet.

He made it below deck to do his rounds and check in with everyone before he went to his own captain's quarters. He had to update his logbook.

June 30, 1834

A storm took us by surprise. By some miracle, the ship took minor damage. We were blown off course. I can't yet determine where we are. We lost four men. Kaufer, Ashburn, Mattocks, and Gally. They were good men.

Cargo is undamaged. Passengers are secure.

The clouds are too thick to see the sun. The waves are calmer, but the air still smells of rain. It is hard to determine if the storm will return.

Repairs are being made promptly. I will set a course as soon as we determine our position.

He sat back in his chair and rubbed the back of his throbbing head. He was still in shock. He couldn't believe what they had experienced, or that they lost men. In all his years on the sea, he had never seen such a storm, nor had he lost men to one.

There was a knock on his door.

"Enter," he called out to them.

The door opened and one of the passengers entered. It was the same one he had spoken to about his daughter. The man had thicker facial hair now, which was to be expected after their long travel. He still looked shaken. Edmond didn't blame him. If that was his first storm at sea experience, then it would definitely be one he would never forget.

"Captain Conway," the man's voice shook. "I was glad to hear that you were okay."

"As I for you," Edmond affirmed. Although, he didn't mention his concern for the salary.

"Do you know when we will make it to our destination?" the man asked. "Time is sensitive."

The captain felt a bit of anger bubble under his skin. Four of his men were dead and all this guy cared about was his work. He had to let that fire inside him simmer down. After all, this entire trip was a business affair for him, nothing more. Edmond knew that the passengers wouldn't care about the losses and damages; they would only care about meeting their deadlines. Men of science always seemed to be more focused on their work than the feelings of others.

Edmond didn't want to chew his head off. He had to keep his cool. He decided just to focus on the topic at hand.

"We are off course," Edmond told him bluntly. He hadn't yet been able to determine their location. "Therefore, I can't give you a real estimated time of arrival."

"I see." He rubbed his arm roughly, as if he felt a sudden chill. "When will you know?"

Straight to the point. Typical.

Edmond could be straight to the point as well. "I don't know," he said honestly. "I will need to work with navigation." His head was killing him to the point he didn't even want to think about it right then. He wanted a cool rag press, something to take for the pain, and maybe another nap.

"When will you do that?"

This man was getting on his nerves. He didn't want to rip his head off—he really didn't. There were many things he was trying to handle and resolve all at once. "It's being resolved," he said shortly. "I'll need time. There are many things. Please give me time."

"Okay." The man nodded. "Thank you for your time, Captain Conway." He left the room after he dipped his head. Probably going after more fish, knowing him.

Edmond didn't want to move, but he needed to figure out where they were. They needed to get back on course.

Chapter 5

"So... we're... lost?" Edmond asked Willas. Both of them stood over the map with a broken compass.

"Looks that way. For now, anyway." Willas shrugged. "Without the sun or the stars, there isn't much we can do."

"The crew won't like to hear it. Neither will our guests." Edmond sighed in frustration. "The clouds aren't going anywhere, it seems. It could be clear in a day or so. Until then..." Edmond didn't finish. He didn't have to. With the sails still in disrepair, they weren't really going anywhere for the time being.

"They'll clear." Willas gave his captain a nod. "Then we will get right back on track."

The crew was already in low spirits after the tragedy. They weren't given any reprieve to hear they were basically dead on the water. They gathered the crew together to break the news.

"How long will this take?" the researchers asked among the crowd that stood on the deck. Both of them looked to be nervous wrecks. Whether they were scared from the storm or about potentially compromising their work, it was hard to tell.

"A day. Maybe more," Willas spoke from next to his captain. "We will see this out as it unfolds. Night will be falling soon. We will need to be ever vigilant in hopes another storm doesn't approach." The crew went into a bit of an uproar at the words, all terrified to face such a nightmare again. They fought about how to best secure everything and who would be keeping watch.

"Simmer down," their captain spoke to them now. "We will work as a crew, as we always have. The worst is surely over by now. After these clouds clear, it is business as usual. Now, who is ready for dinner?"

DeVites took that as his sign to head below deck to start cooking—a good call. Some warm food in their bellies may just smooth out an ounce of the irritation that flowed in everyone.

Everyone took their portions while the researchers took a little more. DeVites was annoyed, but didn't bother to take it to the captain that night. There were bigger fish to fry. They could deal with their greediness at a later date, if at all. If luck had anything to do with it, they would be dropping these men off soon enough. Who knew how far off course they had been taken by the storm?

Night closed in around the ship, and it seemed darker than usual. No moon, no stars, no source of light except the lanterns they had lit about the deck. Edmond paced the deck, looking out into the darkness. It was eerie, like staring out into a hollow void of nothing. He could see the water below and feel it rocking the ship. It looked as black as ink. Beyond what was right there next to him, he saw nothing. He couldn't remember ever being in such a darkness as this before.

The wind swept through his hair, and with it came a tune. He turned his head back towards the darkness. He could swear what he was hearing was… singing? A woman singing, at that. It was just barely audible over the waves and wind, but the tune carried well enough that he could hear words. It was too far away over the waves to make out the words, but they were words nonetheless.

Was he hearing things?

Maryann.

Was he hearing his sweet Maryann?

Just then, Willas was there next to him, shining a lantern out in front of him. All they could see was the reflection of the small flame on the ripples of the black water. "Tell me you hear that singing, too?" Willas's eyebrows were creased in either concern or confusion.

"I do," Edmond confirmed. He thought he was the crazy one, hearing it all by himself. Knowing that he wasn't alone was a relief. Or was it? It meant he wasn't insane, but it also meant that there was indeed a woman somewhere out in the distance, singing a song ever so softly.

"What's that?" DeVitas came up behind them.

"Don't know." The captain squinted his eyes in the dark, only to still see nothing.

"Maybe the wind?" Willas offered. It was a solution, albeit a weak one.

"That's no wind." Halcreek paused along his deck stroll to join them. He looked about as pale as a ghost. "That's evil incarnate."

"Oh, come off it," DeVitas scoffed. "Women aren't all evil." He offered a laugh as he looked to the rest of them. No one else was smiling. The voice had stopped singing by now, but the sound still was fresh to their ears.

"That ain't no woman either." Halcreek shook his head. "But it sure knows know to mimic one."

A wave of uneasiness washed over Edmond. He had heard tales of these things before. Every sailor had. What were they called? Sirens? Merfolk? They were creatures that imitated the beauty and likenesses of women in order to lure men to their deaths, but they were just folklore. Legends. Tall tales. They were stories lonely men told at sea when they were missing the touch of a woman. They weren't real. Surely, Halcreek wasn't suggesting so.

"More likely the wind," Willas offered again, but he couldn't hide the goosebumps that stretched the length of his arms.

"Best we load the rifles." Halcreek looked to his captain as if waiting for his approval.

"You can't be serious." Willas eyed him. "You don't think?"

"That's a siren," Halcreek confirmed what Edmond had already been thinking.

"Grow up." Willas rolled his eyes.

"You hear it, too. You know that isn't the wind," Halcreek snapped. It would have been almost silly to have been coming from any other man, but Halcreek? He was a learned man of medicine. He wasn't one to give in easily to spooky children's tales. The notion of it sat heavily on Edmond. He wouldn't admit it out loud, but he could admit to himself that he did feel a longing for the voice, as he thought it was his Maryann at first.

"The rifles," Halcreek said again. "You'll be praying for them."

"How is Piles?" Edmond broke the tension with his question. He had gone to check on Piles, but the man had been asleep.

"Not good." Halcreek thankfully took the bait and focused on this new topic. "Most of my supplies were shattered and broken during the storm. I don't have anything to clean the wound. I did my best to reset what I could, but he needs a real hospital. It is likely he'll lose the leg or never be able to walk again if he doesn't get help soon.

Grim news. Edmond was hoping for better. He could only imagine the fear and pain that Piles was going through. Hopefully, sleep would do him well.

"Is it… getting foggy?" Willas asked. Where once the lantern only reflected off the water, now it cast a reflection on a cloud of fog that seemed to be growing thicker by the minute.

They could all see it. There was no denying it. It enveloped the ship, rolling across the deck, like a cloud itself had come to rest right over them. They could only see each other and a few feet in front of them. While the lantern was nice to have on a dark night, the fog only reflected the light back at them, making it harder to see beyond the length of their arms.

"They bring the fog." Halcreek looked even more terrified now. "It's a trick. They mean to separate and blind us."

"Would you stop that already?" Willas snapped at the doctor. "The crew has been through enough; it doesn't need you spouting off horror stories."

"Mark my words!" Halcreek began to storm off. "I'll be hugging a rifle below deck if anyone needs me." It only took a matter of seconds for him to be swallowed by the fog.

"Coward." Willas shook his head.

Edmond didn't know what to think. He was sure to side with Willas on this, for the sake of sanity, but at the same time, he knew what he felt when he heard the song. He was feeling a pull; a lull that called him forward. Had Willas not startled him, he wondered if he would have truly cast himself into the sea to see his wife again.

Chapter 6

July 6, 1834

The clouds never leave. The rain comes and goes. It is as if we are stuck in an eternal storm. I hear thunder in the distance even now.

Holes have been patched and sewn. We are capable of travel, all the while having no discernable direction to go. The sails remain closed. We drift about as we wait for the sun or stars to return.

Night brings the singing. Singing brings the fog. The fog is so thick that you can barely see your hand at the end of your outstretched arm.

Halcreek insists it is sirens. He swears we are all doomed. I try to keep the morale up among the men. They are agitated and short. They want to finish the job. We didn't plan to be stuck here for a week. Heaven forbid we remain stranded for longer. Our food will dwindle, the men will grow restless. We will need to fish in order to get more food. We have the nets. Now, I just need to tell the men this without them taking this news to panic.

Edmond Conway rubbed his temples, needing to stop writing to rest his head. It was late and he was ready for bed. He was getting used to the nonstop headache he seemed to be stuck with. The pain was a dull hum, constantly reminding him of the incident that brought them here.

Tomorrow, he would need to begin fishing. The researchers had brought much food with them for the journey, but they were eating it so fast, it hardly made any difference.

He put out the lantern and felt his way to his bed. He climbed onto his bed and laid back. He usually let the rocking of the ship soothe his troubled mind, but tonight, it wasn't working. He was thinking too much about everything. About the absurd circumstances surrounding them as well as this delay keeping him from home longer. He let out a long and tired sigh. He wondered what Maryann was doing right now. If she was sleeping peacefully or if she was out on the harbor with her nightgown blowing in the wind. If only he could hold her right now. It would do them both some good.

As he rested his eyes, his ears picked up on something.

A soft… whimper?

A weeping?

A woman?

For half a second, he thought that perhaps it was only his mind, imagining how sad Maryann would be to see how late he was returning. He knew she would be worried sick at night while lying alone in their bed.

Edmond sat up to listen more intently.

There it was.

He could hear it still. Muffled. Coming from… somewhere.

He quickly climbed out of bed and lit his lantern once more. He held it before him and paced about the room to try to get an indication of where it was coming from. He felt mad to be hearing it at all.

It was coming from outside of his room.

He walked about, following the sounds of the crying. On occasion, it would stop abruptly, only to start up again after a short pause. He checked on his men. All of them below deck were asleep—not that he figured it was one of them anyway. He left them to sleep and continued his hunt.

The sound led him to the cargo hold. There, he heard it clear as day. There was definitely a weeping woman somewhere nearby. He was concerned for a moment about some sort of stowaway, but how could a woman have snuck on board with none of the men seeing her? It would have been impossible.

The weeping was soft and weak, yet it seemed to echo off the boards around him. It was eerie, in a sense. He shined the lantern around to inspect each crack and crevice, only to find nothing.

"Hello?" Edmond called out. There was no way he could have missed her. He searched everywhere. The longer he searched and the more he listened, the more unsettled he became. Her cry sounded… wrong. It was less of a natural cry and more like some kind of a recording playing in a loop. It was the same cry over and over. A weep, followed by a gasp, followed by a soft wail, followed by a whimper, then silence—constantly on repeat. It made his skin prickle with goosebumps.

The crying stopped again. He narrowed his eyes in the dark to strain them further. "Hello?" he called again. "Don't be afraid. I won't hurt you."

Nothing.

No response.

"You okay?"

A voice from behind the captain made him jump. He flung around to shine the light behind him. It was one of the researchers, the same one he always spoke to. He assumed the other one didn't speak as much English as this one. He only wished he could have remembered his name. At this point, it was too awkward to ask again.

"You only scared the life out of me," Edmond said in a winded gasp. "What on God's Earth are you doing down here so late at night?"

"I thought I heard something," the man said with a serious tone. "You?"

The captain took a second to think. He wasn't sure if he would sound like a madman. Did the researcher hear it, too?

"What did you hear?" Edmond decided to ask him.

The researcher shrugged. "Hard to say. Sounded… unnatural." Maybe he had heard it, too. Either way, the captain wasn't sure what to think. The man let out an awkward laugh. "Emmalien's Constellation isn't haunted, is it?"

Edmond didn't find the notion to be funny at all. It was as if the researcher were calling the ship cursed or something.

"No," he reassured the man.

"Captain!" he heard someone calling out above deck. "Captain!"

He could hear a lot of commotion from outside. He rushed past the researcher and hurried above deck to meet Keller.

"Captain!" Keller closed the gap between them.

"What's wrong?" Edmond asked, following Keller as he waved his hand, indicating he wanted Edmond to follow him.

"It's Davis! He went overboard!" Keller said in a panic. Edmond saw that Willas was standing with Bell at the side of the ship. Both of them were struggling with a roped net, screaming down below for Davis to grab on.

Edmond reached the edge of the ship to see what he could do to help. Then, he heard the singing. The echoed voice traveled across the waves. Edmond saw Davis. He wasn't trying to grab the rope. He seemed in some sort of daze. He merely bobbed up and down, looking away from them.

"Davis!" Edmond yelled in his most commanding voice, "Grab the rope!"

He wasn't sure what he expected. The man ignored him, too; bobbing about with the waves, staring out into the inky blackness.

"We'll have to go down and get him," Edmond insisted. He immediately took off his jacket and socks. He meant to retrieve the man himself. By the time he grabbed the rope to climb down, he took one last look at Davis, just to see him disappear under the surface. It was as if he was violently yanked down. The man didn't even have time to scream.

"Davis!" Willas called out to no avail.

The men scanned the water. It was too dark to see anything. If Davis managed to resurface too far out, they would have no way of seeing him.

"Davis!" they each took their turn to yell for the man, trying to give him a chance to scream back in return.

"Was it a shark?" Bell asked. "You think something got him?"

"Davis!" Edmond didn't want to give up. He still paced the deck, looking for anywhere the man could pop up at.

Suddenly, a scream could be heard.

An ungodly scream.

A scream of a man in undeniable pain.

They looked to where he had disappeared but saw nothing.

The scream carried over the deck. He must have somehow wound up on the other side of the ship. The men rushed to the other side and saw Davis. He clawed at the side of the ship, trying to climb his way in.

"The rope!" Keller yelled. The men worked together to get the net to the other side of the ship. They dropped it down next to him in the black water. He managed to clutch his hands around the rope so tightly that he might as well have had talons instead of hands.

He kicked and screamed his lungs out as they pulled to lift him from the water and back onto the ship.

He landed on the deck; his ribcage slamming into the wood with a sickening wet thud. He never stopped screaming. He flung his arms and legs around wildly.

"Davis!" Edmond hunched over him, "Davis, talk to me!"

He grabbed the man's shoulders and rotated him over, all while trying not to be hit by how wildly he was swinging around.

After he was rotated, the sight took all the men by surprise. Davis was still screaming, as any of them would be, for he was missing his eyes and his lips. They were gouged right out of his head and ripped right off his face. The hollowed-out holes in his face were still bleeding. His exposed teeth stretched out in his continued scream.

"Get… Get Halcreek!" Edmond suddenly screamed to the men. "Someone get the doctor!"

Willas somehow tore himself from the horror to fetch Halcreek as quickly as he could.

Edmond grabbed hold of Davis to try and still him. The man was thrashing about to a point that Edmond was worried he would harm himself further. Edmond sat on the deck and pulled Davis between his legs to wrap his arms and legs around him. He had to restrain and calm him.

"It's okay, Davis. Halcreek is coming! You're okay now!" Edmond tried his best to deescalate the terror his man was feeling.

Davis was shouting something, but without lips, there was no way to determine what he was trying to say. Whatever it was, he was screaming it repeatedly. It was something that sounded dangerously like, "Devil."

Chapter 7

The next morning, Edmond sat at his desk across from Halcreek. Willas paced the floor by the wide window behind the captain.

Halcreek looked pale, with dark circles under his tired eyes. He cleared his throat and frowned. The men were sitting in silence. None of them knew what to say at the moment. Each of them wanted to find some solution, but each of them were clueless about what to do.

"Will he die?" Edmond finally broke the silence, not able to look Halcreek in the eyes. The topic too grim.

"I don't know," Halcreek said softly. "The bleeding stopped, but the screaming didn't. He screamed all through the night until he managed to pass out."

"What happened to him?" the captain asked, but it wasn't a question directed toward anyone. It was more of a question he simply put out into the universe, hoping for an answer.

"Sirens," Halcreek still stood by his original statement.

"We're still talking about children's tales?" Willas stopped pacing to glare at Halcreek. "He was attacked, Halcreek. He was almost eaten, and you want to go on a rant about—"

"You heard the singing," Edmond spoke up to silence Willas. "Davis was in a trance—you must have seen it."

"You are seriously listening to Halcreek on this?" Willas seemed utterly shocked.

"I don't know," Edmond said, feeling torn. "I can't say anything with certainty. I saw it all the same as you. What we believe doesn't matter. The facts are, we lost four men while another two men are fighting for their lives."

"We need to leave this place," Halcreek told them. His voice was still steady, though he must have been exhausted. "Let loose the sails. Let's go. The direction doesn't matter; we just need to put as much distance between ourselves and this place as possible."

"We don't even know what '*this place*' is!" Willas stormed to the table to look at the map. "What was the course the researchers gave us? Where did they want us to go?"

"I haven't taken this course before." Edmond pointed to the map. "They said they knew a shortcut and for us to follow this route." He shrugged. "They couldn't have predicted the storm."

"Which storm?" Willas scoffed. "We've been in a storm for a week already. It's not natural."

"It's their storm," Halcreek told them. "We aren't lost. I know exactly where we are. We're right in the middle of their home."

"I respect you, Halcreek," Willas told the man bitterly, "but I can't abide with this nonsense."

"You are in denial. You're afraid." Halcreek tried to say in an understanding tone, "I am, too. Which is why we need to leave. Now. Before nightfall."

"Alright." Edmond nodded. "Alright." He was more than willing now, if nothing else, just to return home to get his men some help. "Tell the men. Let loose the sails. We'll leave immediately."

"Thank you!" Halcreek stood, his face finally showing some emotion. "I'll handle it!" He left the captain's quarters in a hurry.

Willas watched him leave before he spoke to Edmond. "You don't really think…?"

"We need to get the men to safety," Edmond told him flatly. "We'll sail today, then tomorrow, we will take time to fish."

The sails clapped in the wind. Thunder and lightning loomed behind the clouds above as the ship set sail at last. If nothing else, they could sail until they reached the edge of the storm. *Anything for clearer skies.*

At the helm, Edmond felt a sense of urgency he hadn't before. It was hard to place. He wasn't sure if he was running towards something or trying desperately to run away from something.

The waves caressing the bow again was a refreshing sound. They were back on the move.

That was, until the fog started to roll in.

Starting out thin, it didn't take long before the fog was so dense he could barely see a few feet in front of his own face. Without seeing ahead of him, there was no way to know if or when they could dash the ship along a rock side. It wasn't likely they were near land, but he couldn't take the risk—not with Emmalien's Constellation.

The men knew the routine. It wasn't long before the sails needed to be drawn up again, just when they were trying to finally leave. It was as if something didn't want them to go.

Halcreek fumbled up to the helm to find Edmond. He looked fearful.

"We must keep going! We have to!" he declared.

Edmond wanted to as badly as Halcreek did. He needed to be away from this place. But what was he to do? He was blinded by this fog. "When it clears, and not a second later," Edmond assured him, "we will make it home."

Over the creaking of the ship and the splash of the waves came the singing again. It started soft and distant but quickly grew in volume, getting ever closer.

"They'll keep going until each of us walks right off this ship ourselves."

"What do you propose we do?" Edmond gave in to the idea. He had sat on the fence for long enough. After seeing everything he had seen on this trip, he was more than willing to believe in devils and sirens now.

"We sail. Regardless of the fog, we sail onward!" Halcreek spoke over a low rumble of thunder.

"I wish I could," Edmond said sadly, "but I'd put the whole crew at risk."

"They're already at risk," Halcreek persisted.

Edmond's hands were tied and his foot was down. They couldn't do this. Not with his ship. Not with his crew.

A sudden screaming was heard from across the deck. Both the men turned their heads, then looked back to each other. Halcreek had a very "told you so" expression on his worn face.

Both of them made their way down from the helm to investigate the sound. By then, there was more than a ruckus. Following the sound, they came to discover Davis. He had managed to scramble his blind way up to the deck, screaming like a madman. His eyes were bandaged, but his mouth was still in ghastly condition and opened like some sort of ghoul.

"Take him back below," Edmond shouted over the screams to Halcreek.

But Davis wasn't so easy to handle. He flung around and fought madly. What he was doing, Edmond had no idea. What on Earth possessed him to come up to the deck?

Halcreek got a grip on him and tried to wrestle him without hurting him. "Davis! Davis, it's me, Halcreek! Calm down!"

The singing.

It was the singing that was driving him mad. Edmond looked off towards the fog all around them. Something out there was calling for Davis. Whether Davis was willing to fight his way to get to them or he was terrified, Edmond couldn't tell. All he wanted to do was get Davis back into a cot for his own safety as well as for the others.

Halcreek did his best to drag Davis step by step, but it was clear he needed help. Bell stepped in to get a grip on Davis's legs. Together, they managed to take Davis back down. Edmond considered telling the men to gag the poor man while they were at it. His screams were haunting, but he couldn't do it. Davis deserved better.

Edmond took a moment to breathe, feeling torn. He felt he was letting his men down all while trying to keep them safe. Was he supposed to run? Would that help, even through the fog? It was only now that he realized the singing had gone quiet.

He slowly and carefully made his way back up to the helm. Upon approaching, he noticed something on the deck. Something… odd. He couldn't quite make it out through the fog.

He closed the distance by the wheel and finally came to a stop, the object at his feet.

His heart was ice in his chest.

There, with the tendrils still attached and still damp from the sea, were Davis's eyes.

Chapter 8

July 9, 1834

We lost both Bell and Shepherd last night to the sea-to the devils of the ocean. How? I cannot say. They were on patrol last night. This morning, they were simply gone. Nowhere to be found on the ship. How could I be so foolish? Only two men to patrol at night? Maybe in average times, but these are not average times. I practically sent those men to their deaths. It is probably as Halcreek says. Will we simply plunge ourselves into the sea one by one? What could have cast this curse upon us?

Piles gets worse by the day. His infection will kill him at this rate. Davis is still as mad as a march hare. He doesn't eat. He barely sleeps.

Speaking of food, we run low. The researchers demand more than their fair share. I don't allow it. We've already run out of wine. We are low on water. I must reduce the rations.

We didn't have the bait for fishing, but still made the attempt. Casting out our nets, we had several pulls, only to draw them in and find them shredded. Do they mean to starve us? Drown us? Mutilate us?

On that note, Willas believes now. After I showed him the eyes, he had no explanation. None of us did, except for monsters. Monsters that we now know can climb onto the ship any time they wish.

Edmond barely slept that night, tossing and turning. Hearing the ghostly weeping through the walls didn't help at all. He didn't bother to investigate; he was far too tired, even if his mind wouldn't allow him to sleep.

The morning came with a knock on his door. He had barely slept a wink. "Enter," he called out.

It was Halcreek. His expression was solemn. "Did I wake you?" he asked as Edmond was pulling his worn body from the sheets.

"No, no." He shook his head. "Tell me. What news?"

Halcreek frowned. "Well. Two things, and you aren't going to like either of them." His expression alone didn't sit well with Edmond. He could feel the dread tightening around his throat. He waited for Halcreek to finish. Instead, Halcreek only insisted for Edmond to follow him.

Within his makeshift sick bay, DeVitas and Willas both held Davis down onto the cot. It was no surprise that he was still throwing his mad tantrum.

Finally, Halcreek told his plan to Edmond: "I don't say this lightly, but we need to move Davis to the brig."

"The holding cell?" Edmond was shocked. "The man's a victim! A patient! Not some criminal!"

"A violent patient," Willas spoke up, clearly struggling against him. "One who means to throw himself back into the sea."

"We can't." Edmond couldn't stand the idea of it. "He is injured. We can't lock him up like some kind of rabid animal."

"It's us, or him," DeVitas said. "You want him to hurt someone else? Himself? It's for his own good, Captain."

As much as Edmond hated it, he knew they were right. Trying to tend to Davis like this just wasn't practical, and restraining him to a cot wasn't ethical. The brig was the best and safest option for everyone.

Edmond finally gave a small nod. That was enough for the men to make quick work of it. Edmond followed the three of them as they walked Davis to the holding cell.

Once they locked him within, the guilt and shame hit Edmond the hardest. The sight of Davis walking around blindly in the room with his arms out made Edmond uncomfortable. This was Davis: a good man. A man who had trusted Edmond as his captain to lead and protect him. Yet, Edmond had failed him—failed him and betrayed him by locking him up. He needed help, not prison, but there was only so much that they could do.

This was for the best. They would make it home and be able to get him real help. Davis wasn't himself right now; he needed this confinement. At least, that is what Edmond kept trying to convince himself of.

"That's one," Edmond said sadly, watching Davis trip over his own feet and babble incoherently.

"What?" Halcreek sounded confused.

"You said there would be two things," Edmond reminded him. "This was one."

"Right." He winced at the realization. "You really aren't going to like this next one, but…" He paused as he thought about how to break the news. "It's Piles. That leg. It has to come off."

"You're kidding." Edmond's heart dropped into his stomach. "Surely, there's another way."

"It's badly infected. If it doesn't come off, the infection can spread. We need to amputate."

They began to walk back to the sick bay while Halcreek continued talking. "I've got nothing to give him for the pain. Therefore, it will be a very messy operation. Having you there may help give him strength. Plus, I may need your actual strength to hold him down. It won't be easy."

Willas joined them, the three of them standing over Piles.

While Halcreek prepared, Edmond looked Piles over. He hadn't noticed just how bad he had gotten. He looked horrible. He was covered in a layer of sweat so thick that even his dark brown hair looked black from all the moisture. Meanwhile, his typical caramel skin was pale and ghostly. His limbs were trembling, and his teeth chattered. His leg looked grotesque—dark purple with the shard of broken bone still sticking out. The wound was black around the bone.

"You'll want this, my friend." Halcreek had a wad of cloth he meant for Piles to bite down on. Piles gulped audibly before nodding and opening his mouth to clamp down.

Edmond felt horrible for Piles. For him to know what was coming and needing to accept it, knowing that it was either this or potentially dying... Having to make such a traumatic and painful choice all without an actual choice…

"I'll need to take it up to here," Halcreek spoke to Piles while he pointed just above his knee. "Once the infection is removed, I will clean and dress your wound. This will hopefully prevent a new infection."

All Piles could do was nod, but Edmond saw the single tear that trailed down the side of his face and into his hairline.

"We are here for you," Edmond felt the need to remind Piles. He held his hand with one hand and held his shoulder down with the other. Willas sat right on the other side to hold his other hand. Meanwhile, Piles had his free leg strapped down.

Halcreek took out the saw and Edmond's stomach turned. He didn't want to be here for this—not at all—but it was his duty. He was obligated to be here, but his stomach wasn't built for this. Seeing the saw alone made Edmond feel a phantom of the sharp teeth in his own skin. He looked back up to Piles's eyes.

"Are you ready?" Halcreek asked everyone. Piles nodded and winced hard, keeping his eyes closed.

Halcreek nodded to Willas and Edmond, and they tightened their grips.

That was when he lowered the saw and let the teeth sink into Piles's skin. The very first drag of the blade caused Piles to throw his head back, screaming into the cloth. His grip on Edmond's hand was beastly. At that moment, a flash of Maryann came to Edmond's mind. Specifically, when he held her hand during her first delivery. He could almost see the way she writhed in pain while she leaked blood, sweat, and tears to bring Emmalien into the world. His mind traveled to the new baby—to the delivery he needed to make it back home for.

A wet squelching sound brought Edmond out of his daydream and back into the now. He shouldn't have, but as if by reflex in reaction to the sound, Edmond looked down toward Piles's leg.

Blood flooded the cot beneath him while the saw buried half-way into his leg. The saw was smeared with blood and chunks of severed skin. The cut wasn't exactly clean.

Beyond the squelching, the saw blade made a grinding sound. It had met bone. The teeth dragged hard back and forth, but they were making slow work through the bone. Each tug and push of the saw blade sent jolts of reactions through Piles. He began slamming his head back onto the cot repeatedly. His chest jerked and rose quickly with each pain-filled scream he let out. Blood sprayed out, pulsating in a thin stream onto Halcreek in rhythm with Piles's heartbeat.

As if he wasn't sweating enough, Piles was now so drenched that Edmond was almost losing his grip on the man.

The rough grinding was followed by a sickening snap that made Piles's eyes roll in the back of his head. Then, he was out. The pain must have caused him to finally black out. He had had all he could take.

Edmond didn't dare watch anymore. He kept his eyes shut and kept his grip on Piles while he listened to the saw finish squishing and gnawing the rest of its way through the back of Piles's leg.

He could hear Halcreek moving. He heard him rustling things around and going about his work quickly, but still, he refused to look. He didn't want to see it. He couldn't bear to.

It wasn't until seconds, minutes, or hours had passed that Halcreek was finished. "It is done," he told them.

Willas let go, looking pale and sweaty himself.

Edmond released Piles. His fingers and arms were aching from how hard of a grip he had had. He glanced down to the leg only to find a stump dressed in bloodied bandages. He didn't see the lower leg, and he was thankful for it. The only other sign of this incident was the blood that covered the cot and Halcreek himself.

"You can go now," Halcreek told them both. "I can handle the rest."

Edmond didn't need to be told twice. He nodded quietly and slowly, as if in some kind of daze, and walked from the room. His mouth was as dry as cotton, his throat was sore, and his stomach was in tangled knots. He needed fresh air.

In slow motion, it seemed, he made his way up to the deck. The breeze felt refreshing on his face. He definitely preferred the smell of the rain over the coppery smell of blood.

Blood.

He had seen too much of it today.

Enough to last a lifetime.

He approached the side of the ship, looking at the water below. The waves lined with white as they hit the side of the hull. This time, his stomach rocked right along with the ship. He leaned over the side and promptly began to vomit.

Chapter 9

That night, the captain called everyone to the deck for a meeting. Everyone except for Piles and Davis, that is.

The crew, once hardened men, looked a bit shaken. They were hungry. They were tired. They were confused and afraid. Edmond had to keep a brave face for the men.

The fog hadn't lifted, so seeing all the men's faces in the back wasn't easy. He wondered if they could all see him. Probably not, but as long as they could hear him, that was what was most important.

Edmond began, "You all know about Bell and Shepherd. It was a tragic loss, and it could have happened to any of us. Therefore, we need to be more mindful."

"How could they just fall into the sea?" Marlin asked. "It just doesn't make sense."

Of all the things Edmond had witnessed, he wasn't sure how much to share with the men. Would he even be believed? They knew Davis had been attacked, but they didn't know by what. The men didn't know about the incident with Davis's eyes, either—only Edmond, Halcreek, and Willas were in on that much.

As much as Edmond wanted to protect his men, he also didn't want to spook them unnecessarily. But was this really unnecessary? How much could he share that could potentially save their lives?

"There's no saying," Edmond answered while looking at Halcreek. He didn't want him to start yelling all around the deck about monsters in the sea—at least, not in his over-the-top manner. This was a delicate matter.

Edmond continued, "We will need to work in groups of four or more. Make sure that in all you do, you are with others. There is safety in numbers."

"What about rifles?" Halcreek asked. It was as if he was just waiting for his moment to say something.

"Why do we need rifles?" the researcher spoke for some of the men. "What are we shooting?"

"We don't have enough rifles for all the men," Edmond reasoned, even though he did see the potential need for it. "All who patrol in the night will have them."

The researcher felt the need to press, "What are the rifles for?"

Edmond sighed. Would he believe him? Would the whole crew find him mad? Would they think he had lost his mind?

"For protection," Edmond told him. He turned and nodded to Halcreek. The men looked at each other, wondering who to group with.

"Tonight, I will patrol." Edmond felt it was his duty to protect them, and that meant putting himself in front of his men. "Which three of you will patrol with me?"

"Me!" Halcreek was quick to raise his hand. He seemed happy to finally be getting his rifle.

"Aye!" DeVitas volunteered.

"I'm at your side. Always." Willas clasped the captain's shoulder.

The rest of the men went below deck to try their hardest to get some sleep.

The four men collected their rifles.

The night seemed calm. No rain, no wind, no thunder, only fog and the creaking of the ship. The four of them stayed evenly spaced and spread out along the ship. They couldn't see each other, but called out to each other every few minutes. If anything climbed onto the ship, they would be sure to see it.

There was a whisper over the waves that came soothingly to Edmond's ear. At least, it started as a whisper. Then, it slowly grew into a soft song. That beautiful voice again... He swore it sounded just like Maryann.

He could actually hear the words now. He could make out what her voice was singing.

Sink within below the waves
The surface never knew you
A peace that everybody craves
Surrounded by the vast blue

Sink within as you descend
Ears muffled by the deep
No more needing to pretend
Finally, you can sleep

He paced to the edge of the ship, looking out into the night. He hadn't expected to see much. He never did. The song continued.

Sink within and find your peace
Let the waters fill you
The burning in your lungs will cease
And all the pain you've been through

Sink within this cold embrace
You've found where you belong
Your vision dims of this last place
For now, it won't be long

He felt a tugging at his heart—a tugging at his whole body. He glanced down at the water. It looked so warm, so inviting, like it was welcoming him to rest.

Sink within the cool allure
Your home is the depths below
You'll never be more secure
Become undone and swallowed whole

Sink within the endless void
The dark and stillness you pursue
A shell to a soul has been destroyed
The surface never knew you

He hadn't realized over the course of the song that the voice was growing louder, closer. It was what his ears heard next that caused him to freeze. A wet slapping sound, like wet cloth hitting the deck repeatedly.

It came from his right, back up towards the helm. He raised his rifle and slowly approached through the fog to see the back of a woman. She stood at the back of the ship, looking out over the sea.

Edmond raised his rifle.

She must have heard it.

She turned around to face him.

His heart clutched tightly in his chest.

It was Maryann. His sweet Maryann. She wore her white nightgown that almost blended into the fog. Her auburn hair fell over her shoulders in curls. She rested her hand on her very swollen stomach.

"My love?" She looked frightened.

"Maryann..." He lowered the rifle, his voice breathless. "My love, what are you doing here? How?"

"Come." She lifted her free hand to him, her delicate, slim fingers reaching out for him. "Come to me, darling. Hold me. Kiss me."

His heart fluttered in his chest. Was he going crazy? He missed her so much. He wanted nothing more than to take her into his arms.

"My sweet, please," she said softly, "I'm so cold."

Sanity be damned, he wanted his wife more than anything now. He dropped the rifle and walked towards her, his arms open to embrace her.

A loud shot cracked the sound barrier and seemed to rattle all of Edmond's organs, making him jolt at the sound. A gun shot, fired so close to him.

He turned around to see who had fired, only to see Willas and the others running towards him, the fog swirling around them. Their mouths moved in completely silent screams—all he heard was her voice.

"My love!"

Another shot, this one from Halcreek, and he didn't miss. The shot ripped into Maryann's shoulder.

Edmond yelled and threw himself in front of his men to stop their madness, but a blood-curdling screech came from the woman behind him.

He looked back over his shoulder to see... something. Whatever it was, it was certainly not his lovely wife. The illusion was broken. In place of Maryann stood an almost human-looking creature. "Stood" being the key word, because her legs weren't legs. She had a long fish tail that trailed behind her like a snake's. She was propped up in that way. Her torso looked the most human, but even then, only barely so as she was covered in scales. Her face was grotesque—it seemed mostly occupied by teeth, jagged and yellow, protruding sharply from her wide, unhinged jaw. Her eyes were large and black. Her "hair" seemed stringy like seaweed. Her nails were long, her fingers bony and sharp.

This was the monster that Halcreek had warned them of. Edmond knew now what the devil looked like.

She grabbed her shoulder and flung herself from the back of the ship before any more shots could be fired.

The sounds of the world seemed to return to Edmond as he finally heard his men's voices.

"What were you thinking?" Willas shook the captain's shoulders. "Are you mad?"

DeVitas came to the edge of the ship to look over it. He held his rifle up, ready to shoot if he saw anything pop up from the surface.

"Do you see?" Halcreek pointed over the edge, looking at Edmond. "Do you see now how the men will leap to their deaths? They are witches of the sea! Casting spells on men to lure them in!"

"I saw Maryann..." Edmond's hands trembled due to fear and agitation. He couldn't believe he thought that hideous beast was his Maryann. What had that thing done to his mind?

"Of course you did!" Halcreek shouted. "It's what I've been trying to tell you this whole time! We need to leave!"

"We almost lost you!" Willas told Edmond. "Rifles are useless when you're under their trance."

"Did you hear it?" Edmond was still in a half daze. "Did you hear the song?"

The men looked among themselves and shook their heads.

"They sang for you," Halcreek told him. "Just like they sang for Davis. Just like they sang for Shepherd and Bell! They give your mind delusions to see whatever you desire most! They have set their sights on you, Captain."

Chapter 10

Edmond sat at his desk, his mind muffled in thoughts. The quill hung limply in his hand over his logbook. Just like every day, he wanted to add to the pages, but this time it felt different. This was the first time he would have a record of seeing the beast himself in the flesh. He almost didn't believe it. Would the whole book be counted as the accounts of an insane man?

He had meant to write last night, but was still processing the information then. This was real. This was somehow real. In the early morning hours, the fog was illuminated and cast a pale light in through his windows.

He began slowly, unsure of each quill stroke:

July 11, 1834

I saw Maryann last night. Rather, that thing took on her form, her voice, I swear, even her scent on the wind. I actually fell for it. I was lured in. Had my men not saved me, I would be lost now.

The fog still lingers. I don't know how they control it, but they do. Did they control the storm that brought us here?

I want to be strong for the men, but I worry my constitution has been compromised. I don't know what is real anymore. None of this should be happening.

I won't make it home for the birth of my baby, if I make it home at all.

Edmond's hand loosened and the quill fell from his grip. He felt a certain burning in his throat and moisture building in his eyes. His heart ached. He should have listened to his wife. No amount of money was worth this. He should be home with her now. She needed him. He needed her. At this rate, he never wanted to step foot on a ship again.

There was a knock on his door.

"Enter," he said reluctantly. He really just wanted to be left alone.

DeVitas entered. He came to the desk, waiting for Edmond to welcome him to take a seat. Edmond nodded.

DeVitas took in a long, slow breath before speaking—indicating that whatever he was about to say was not going to be good. Edmond felt the dread already. He finally spoke to his captain, "We need to fish. Somehow. Some way. We need food."

Edmond knew this already. He was thankful that DeVitas brought this concern to him, but at the same time, what was he supposed to do about it? The beasts had ripped their nets to shreds. They couldn't cast them.

Edmond nodded. "What do you suggest?"

"I don't know, Captain, but we only have enough food for a few days. We won't make it on that. I've rationed the livestock as thinly as I possibly could. Before long, the men will want to start eating each other."

Edmond felt hollow and defeated. Every single man aboard this ship counted on him. He was a failure multiple times over.

That was when Halcreek's words returned to his mind. They needed to move. It didn't matter where—anywhere was better than here.

Edmond considered it within himself. He had wanted to wait for the fog to disappear, but if those things controlled the fog, then it wasn't ever going anywhere. Surely, they were far enough at sea not to hit anything—or anyone else—sailing. He couldn't imagine seeing another ship here.

"What if we left?" Edmond asked out loud, not really expecting an answer, just mostly rambling to himself.

"Through the fog?" DeVitas raised his eyebrows. "You really don't think it will clear?"

"Doesn't look like it." Edmond wrestled the thoughts in his head, torn between the safety of his men one way versus the other. Either way, they were in danger. He leaned heavily on the idea of leaving. After all, it was the lesser of the two evils. "But running into anything else is almost preferred. We need the supplies. We won't find them here."

"Aye," DeVitas seconded. "I support you, Captain. We should sail before I am tempted to fish one of those beasts out of the water to cook it myself."

Edmond released a chuckle. It was the first time he had chuckled in days. His stomach was growling horribly. He had been skipping meals to make sure the men ate enough. He could only imagine being able to skin the thing who mimicked his wife—the thing that dared to warp his mind.

Another knock came.

DeVitas and Edmond both looked up to the door. Edmond called for them to enter as well.

It was the only researcher who really talked to anyone. "Captain Conway," he greeted him.

Edmond wasn't about to fumble the man's name. He only nodded at the man in greeting. The researcher came to stand by his desk.

"I hope I am not interrupting."

"No. Please. What brings you here?" Edmond waved his hand for the man to take the next available seat.

He didn't sit. He seemed in a rush. "This fog isn't going to go away any time soon. I'd like to get us moving if I can. What can I do to help us along here?"

"Can you clear the fog?" Edmond raised an eyebrow.

"I cannot," he said with a sigh. "But neither can I sit here idly. My research demands we move."

"I don't answer to your research." Edmond couldn't stand the way some of these science men operated, acting so removed from reality and entitled to more than their share.

"My research will change the world as we know it," he said bitterly. "Or at least it would, if we ever arrived at our destination."

"I can't control the weather." Edmond hadn't seen the researcher seem so brash before. He must have been as irritated as Edmond was.

"But you control this ship, yes?" He pointed around. "You can act like it. You can get us moving."

DeVitas stood in a jolt to be in the researcher's face. "Don't talk to my captain like that!"

"DeVitas." Edmond indicated for him to sit again. DeVitas lowered himself promptly.

"I mean no disrespect," the researcher backpedaled. "I only mean…" He hesitated before continuing, "My research. It is time sensitive." Edmond had heard this before. Although, he was more concerned about their food and water supply at the moment. That was time sensitive, too.

"I need time to consider the options at hand," Edmond assured the man. "We will either leave when the fog clears or…" He didn't want to say 'when the supplies run out' to him. Instead, he left it hanging.

"I see," he said. "Please, make your decision quickly. For all of us."

That night, Halcreek wanted to volunteer yet again for patrol. Him, Hammond, Keller, and Willas.

Edmond sat below on his bed, thinking.

Thinking mostly of his wife, but also thinking about how easily he had been manipulated by the beast.

He was exhausted, but no matter how hard he tried, he just couldn't sleep. There was far too much on his mind.

Not to mention, the blasted crying. It was back.

He didn't bother to get up and check it this time. No matter how hard he searched, he simply couldn't find a source. Maybe it was all part of the delusion. He wouldn't be bothered by their tricks.

It was a few moments after the crying that he heard something else. Not the singing, as he suspected, but whispers. It was like being in a room full of a hundred people all whispering to each other at once.

It was very faint, but definitely there.

He rose from his bed and lit his lantern. He shined the light at his window to look out, but the glass only reflected the light back into his face.

He would have to go outside.

He went along to the deck. He didn't expect to get much sleep anyway. He found Halcreek looking out over the water with the other three men. They gripped their rifles.

Halcreek spoke to the captain just over the whispers in the air. "Captain, I was just about to retrieve you." He made a nod out towards the sea.

Edmond lifted his lantern and cast his gaze over the black waters that eased and bobbed the ship in place.

At first, he saw nothing.

Straining his eyes, narrowing them.

In time, it became clearer. Just at the edge of the darkness where the fog consumed the world around them, he could see something.

A reflection.

A set of eyes, their face peeking from the surface.

They were being watched.

He moved the lantern, and the light caught another pair of eyes. Then another. Then another.

There were dozens of eyes reflecting back at them from the waters all around the ship. The whispers continued in a language none of them could decipher.

Edmond's heart froze.

One thing was clear: They wouldn't have enough rifle rounds for all of them.

Chapter 11

July 12, 1834

I can't sleep at night between the weeping and the singing. Both are likely to drive me mad, I'm just wondering which it will be first.

Piles seems to be on the mend. I'm relieved by it. Although, Davis still seems to be mad. He sits in the corner of his cell, rocking and mumbling incoherently to himself. Any time we try to approach or offer him food; he lashes out like a wounded animal.

There's no need to hide the truth from the men. They know. They've seen them with their own eyes. The sirens. It is not just one. They surround the ship. They haven't tried climbing onto the ship again. Not since Halcreek shot that one beast. They know we are armed. Instead of attacking, they seem to be waiting. Conspiring. Even biding their time. For what, I don't want to find out.

The fog is ever persistent. They don't want us to leave.

Edmond had had enough as the sun was setting. He put the quill down and stood from his desk. He barely even cared about the salary anymore. He just wanted to go home.

He went above deck to find Keller, Hammond, Shaw, and Halcreek doing their rounds. Halcreek had insisted on doing patrol almost every night. It was as if he held a personal vendetta against these things and meant to shoot every one of them. At least, any that tried to climb the ship again.

"Let down the sails," Edmond said sternly when he approached the men.

"But the fog?" Hammond asked.

"To hell with the fog. To hell with these devils! I'm going home to my wife."

Halcreek didn't need to be told twice. He set about getting the sails opened as soon as he physically could.

Edmond didn't care anymore. He was going to sail. He didn't concern himself with the direction. He didn't concern himself with hitting anything. He only cared about seeing Maryann again.

The sails caught the wind, and they were off. Halcreek finally had a reason to smile. He was finally getting his way. Edmond figured that even if they did hit land, at least they had freed themselves from this loss of direction.

Edmond manned the helm. Feeling the ship glide through the waters again felt right. The rising and dipping of the bow, the spritz of the ocean back on his face, the wind stroking his hair… He was going to continue forward until he found something. Anything.

And sail, they did. They sailed until the sun tried to fight its way back through the clouds only to fail yet again. Willas came to meet him at the helm.

"You haven't slept," Willas said. It wasn't a question. It was evident all over his captain's face. His bags were just another piece of luggage for the voyage home.

"I'm going to get us home," Edmond stated plainly.

"Let me take over," Willas suggested. "You get your sleep. You need it. And we need you to be at your finest."

Edmond trusted Willas to handle the ship while he took some rest. Maybe during the day, he wouldn't be haunted by the cries or the singing. He was sure that his mind was melting from lack of sleep. He could have passed out right there on the deck, but instead, he opted to head below to his quarters.

His bed welcomed him warmly, though his sleep didn't. In his dreams, he held Maryann's hand while she was giving birth. Her hair stuck to the sweat of her face and neck. He helped to hold one of her legs back as she pushed. The midwife coached Maryann through it, telling her when to breathe, when to rest, and when to push. Edmond felt the pressure every time his wife was told to push. Her squeezing his hand made him brace for impact, too. Each push could be the final one.

She screamed in pain. The sheets were saturated in blood. Something was wrong. Maryann clutched at Edmond's shirt, pulling him ever closer to her. "Please!" she screamed at him between breaths, "Please help me!" He felt the fabric tearing from under her grip. She would tear the shirt right off of him at this rate.

"I'm here!" he told her. He let go of her hand to push the hair from her forehead. "I'm right here! I'm not going anywhere!"

"Push!" the midwife instructed.

Maryann let out another scream, her head thrown back as if cursing the very heavens above. She pushed with everything she had, her body tense to the point of trembling.

Finally, the wet squelch of the baby was heard as it was released from her body along with various other liquids. Maryann was still, her head back, her eyes closed.

"Maryann?" Edmond touched her face and squeezed her hand. "Maryann!" He couldn't get her to open her eyes or even respond to his touch. He cupped her face with both hands to give her a gentle kiss before begging her one more time to look at him.

"Your beautiful baby," the midwife bundled the bloodied infant into a blanket.

Blood.

There was far too much blood.

There shouldn't have been this much.

The midwife left bloody handprints on the blanket as she passed the bundle to Edmond.

He released his wife to take his infant into his arms. Pulling the little one close, his fingers brushed the blanket over to check the baby's face.

A horrible monstrosity stared back at him. It had more jagged teeth than he could count, bulbous eyes, no nose, and scaly gray skin.

The weeping.

He heard the woman weeping again.

He looked back to his wife. She was awake now. Her tears were blood that trailed down her face. That same repetitive crying he heard almost every night came from her lips.

"Maryann?" He gasped at seeing her like this. Was this even his sweet Maryann anymore?

Her face darkened and twisted into that of the monster he had seen on the ship that night. Her weeping got louder and louder until it morphed into a hideous cackle.

Edmond flung himself up in bed. His heart pounded so madly, he thought it would burst right through his chest. Sweat drenched his sheets. He took a moment to collect his breathing. What time was it? He wasn't certain—all he knew was that it was night. Again, he had managed to sleep the whole day. He must have been more tired than he thought.

He threw on his clothes and rushed up to the deck. He had to know where they were now.

He arrived only to see that fog still engulfed them all around. He found Willas still in place.

"We've traveled all day?" Edmond was confused to see the surroundings the same as from before.

"We have." Willas looked exhausted now. "Who knows where the end of this fog is?"

"Or when we will run aground."

"Better land than sea as far as I'm concerned." Willas made it clear that he, too, had seen more than enough and wanted out.

"We should have been out of this by now." Edmond just didn't understand.

DeVitas came up to the captain with a bowl of thin soup. "You need to eat something."

"I'm not hungry," Edmond insisted. "I want to know where we are."

A long crack of lightning spit the sky above them, followed by a low rumble of thunder.

Dread washed over the men's faces. They didn't think they would survive another storm—not if it was like the last one.

The clouds opened up to release rain that pattered along the deck.

Edmond sighed in defeat and told his men to raise the sails.

Chapter 12

The rain hit the deck hard. The night was alive with the sharp sound of static in the air. Edmond could barely see through the combination of rain and fog. To open your mouth to talk would mean to spit out a mouthful of rain. They could almost drown by just standing there.

They had barrels sitting open in hopes of collecting as much rainwater as the storm would allow, but the rocking of the ship didn't help.

The ship was driven high and dragged low with each rolling wave under them. While every man feared the rogue wave, they tried to keep their heads about them and maintain the integrity of the ship.

Edmond took over to give Willas a break. The man had been sailing the ship all day. He was drenched all the way through when he maneuvered the rocking ship in order to head below to find his bed. He was better off sleeping than being so tired that he would slip and fall into the waters below, anyhow.

Edmond tried his hardest to keep his eyes on the water all around them, but the rain was too ferocious. He could barely even hear his own thoughts over the sound.

He saw a figure coming to him, a shape moving through the fog and rain. It was Keller. He had to shout at the top of his lungs to be heard over the storm, "It's Halcreek! He needs you below!"

Emond left the helm in his capable hands and pushed through the rain to make it down the stairs. He left a massive puddle trailing behind him. A few times, he was thrown against the wall when the ship would rock violently one way or the other. He pressed his damp hand along the wall as he walked in order to steady himself.

He searched the sick bay, only for Piles, who was hanging on for dear life, to tell him to go and check on Davis. Edmond felt immediate dread. What could possibly be happening with him?

Edmond made it to the holding cell. Inside, he found Halcreek holding a screaming Davis down on the ever-rocking floor.

"I'm here!" Halcreek was shouting over the thunder to Davis. "It's me! Halcreek! Listen to me! You're safe here!"

Davis's face was a bloody mess. His nose was crooked and half hanging off his face altogether. Edmond noticed the bloody smears along the walls. Davis was back to hurting himself. Was the storm provoking this behavior?

"He's going to kill himself!" Halcreek yelled to Edmond.

Edmond came right alongside the men. He wanted so badly to help, he just didn't know how. If only they had a way of putting Davis out—a more ethical way than a knock to the head.

"Let's strap him down." Edmond nodded to the cot in the holding cell. "I'll get the leather." He rushed from the cell to get what he needed outside. Then, together, the men strapped Davis down. It broke Edmond's heart to be treating him like this, but he didn't know what else to do.

"Thank you," Halcreek said a little breathlessly. He spent a good amount of energy fighting a man that seemed to be running on fear and adrenaline. Davis had endless energy to fight and scream; it just wasn't natural.

Edmond clasped Halcreek's shoulder and then stepped out of the room. He had heard enough of the screaming, and he couldn't bear to hear any more.

"Help me!" Upon shutting the door and heading past cargo, Edmond heard a cry for help. A small and shrill voice reached him over the storm. His heart tensed up. Had he really heard that? That was none of his men. It was the voice of a little girl.

Not just any little girl...

Was that... Emmalien?

No. It couldn't be.

Edmond shook his head. The rain was playing tricks on him.

"Daddy!" she cried out to him again.

His head jerked to the left towards the cargo hold. He was certain he heard that as clearly as he had heard the crying most nights. It was real.

He staggered into cargo. Everything was strapped down, so the boxes and containers wouldn't slide or crush anyone. Regardless, he still had to hold onto things while he walked so as not to trip over his own feet in the swaying.

"Emma?" Edmond felt stupid for even daring to call out to his daughter. As if his actual little girl would come running to him like she used to when they played hide and seek together so many years ago. "Emmalien?"

"Daddy, please!" She sounded so afraid. "I'm scared!"

Emmalien hated storms. She was always so scared of thunderstorms that she would demand to sleep in her parents' room at night when there was one. She thought the thunder was actually the growling from a ginormous monster hiding above the clouds—some eldritch horror that brought the storms with it. Maybe he had told her too many sea legends. She would eat them up only to be scared later.

"Baby!" His voice choked up as he called out to her, "Come here! I'll hold you!"

"I'm stuck, Daddy!" she screamed in a panic. "Help me!"

Stuck? He looked around each of the boxes. He had searched the bay so much, he knew every inch of it. That's when he noticed a blanket had been moved off of a large box about the size of a double dresser. A small hand-sized hole was cut into the top. He knew it was hand-sized because out of it stretched a tiny, pale hand just past the wrist.

"Emmalien?" Edmond's mouth was as dry as bone as he stared at the tiny hand in horror. It was as if the rocking of the ship and the whole storm itself ceased to exist. The only thing that was real in this moment was Emmalien.

Her small hand slipped back into the box, and a fraction of her face could be seen. Her mouth at the hole, she cried, "I'm so scared, Daddy!"

Edmond slowly got closer, in utter shock and disbelief. Through this hole, he saw part of his daughter's face bobbing within the black waters of the box. Each sway of the ship brought a wave of water over her pale face.

She's going to drown!

"Emmalien!" Edmond snatched a crowbar from next to the door and ran back to his daughter. He fought to pry the thing open, all while she called out to him for help.

Suddenly, Edmond was tackled to the ground. The crowbar went flying and rang out as it slid loudly across the floor. Edmond wrestled with his attacker, only to see that it was both of the researchers. One sat over him while the other was at the box, checking on Emmalien.

"My daughter!" Edmond yelled to the man as he struggled to get out from under him. "That's my daughter!"

"It's not!" The man forced Edmond's shoulder's down. "That's not your daughter! Emmalien is dead!"

"Daddy?" Her voice was so weak. So scared. So small.

"My…" Edmond's chest went up and down in rugged breaths. "My baby." He was hurt. He was confused. Emmalien wasn't dead. She was right there. His baby was only a couple feet from him, but she was trapped in that box. How did she get inside that box? Who were these men? How did they get Emmalien?

"No!" The researcher shook his head and pointed a finger in the captain's face. "You told me that Emmalien was dead. She died two years ago. Remember?"

Did he remember?

She was so close.

She was right there.

His daughter was just right there…

"My Emma…" he gasped, feeling confusion wash over him.

"Is dead," the researcher repeated. "That isn't your daughter."

The other researcher grabbed a thin sheet and started to spread it over the box.

Edmond looked back towards the hole to see a larger hand reaching out—one that was gray and scaly with long, black nails.

The cloth covered the box, and the hand descended back inside.

It wasn't Emmalien. It wasn't his precious daughter. One of those things had managed to get inside his head again.

The researcher saw that Edmond had returned from his delusion, as he was no longer struggling against him. He stood up to dust himself off.

Edmond backed away, using another crate to pull himself up. While the confusion was clearing from his mind, the shock of it was becoming more powerful. His brain was running at a thousand miles a minute. All the wild thoughts were flashing so quickly in his mind that it was hard to settle on one. Finally, he did. "You brought… one of those things… *onto my ship*?" He shouted the last bit, slamming his fist into a crate at his side.

"It is our research," the man told him. "You will see. This will be worth it in the end."

Edmond was furious. "You—" he sputtered, "That thing!" He ran his fingers through his hair in shock and frustration. "Putting my men in danger!"

"This one isn't violent!" the second researcher spoke up. "She is young. She is what we would consider to be an adolescent."

Everything was beginning to make sense. The weeping didn't work, so she attempted to mimic a child. Edmond's child. His stomach twisted and he felt like throwing up. "You… kidnapped their child…?" He could only imagine what he would do to the people foolish enough to steal his child. It was no wonder they were being stalked and hunted by these things, no wonder they wouldn't let them leave, and no wonder they weren't openly attacking. As far as the sirens were concerned, the men had a hostage—a hostage they didn't even know about.

All of these thoughts and more came to Edmond. What on Earth were they supposed to do?

"She was all we were able to safely secure," the first researcher affirmed. "We never intended to harm her—only study her. We almost made it back, too. I don't know how they found us."

"You put a target right on us," Edmond said through clenched teeth, "You—" He shook his head. "I lost good men because of you! Davis is melting away because of you! We are stuck in this purgatory because of you!"

"We can sail!" the researcher protested. "Just take us to our destination! You'll never see us again after that!"

"We can't sail!" Edmond yelled. "Whether it be fog or storms, they won't let us leave! And now I know why!" He pointed at the box. "Give her back!"

"I can't do that," the researcher said in a voice that seemed too calm for the situation.

"You can, and you will!" Edmond declared. "It's our only hope of being released. I have a pregnant wife at home!"

"Congratulations," the researcher said sarcastically. "You have your hopes and dreams, Captain Conway. We have ours. This project is my baby. That—" he pointed to the box, "is my baby."

"She isn't your baby!"

"Funny," the researcher snapped, "I was just telling you that only moments ago. Do you see why we need to do this? Why we need to study them? Their delusions, their powers. We need answers. We could potentially save so many lives."

Edmond growled. He was angry with the man, but angrier with himself for being fooled. His heart still felt tight and pained from being so close to holding Emmalien again.

Edmond's jaw was clenched. "Get that thing off my ship. Otherwise, the two of you will be thrown into the sea instead."

He rammed the man's shoulder hard as he walked past him to leave the room. He had to tell the others, just as he had to update his logbook. The same book that was currently sounding more like a fairytale than an actual account of reality.

Chapter 13

The researchers stood outside of Davis's holding cell. Edmond couldn't show them the men they had lost, but he could show them the horror they had brought onto Davis. The man hadn't been the same since he had touched the water. They couldn't figure him out. Why couldn't they get Davis to calm down? It was as if every bit of sanity the man had was left behind in the ocean.

The researchers said nothing. Only one of them even bothered to look at Davis. The quieter one only looked at the floor.

"You see what your research has brought onto my men?" Edmond scolded them as if they were children. He only hoped they had a heart enough to care. Davis had a wife. She would be devastated to see what was left of her husband.

"We can't get him proper treatment until we leave here, which can only be done once you release that siren," Edmond told them. They would need a plan to get her out of the box and back into the water without the men coming to further harm. Who knew just how docile she really was?

"That won't help him," the researcher scoffed. He sounded half heartless as he shook his head.

Edmond didn't like the sound of that. "What do you mean?"

The researcher frowned then nodded toward Davis. "Can't you see? He has been kissed by a siren."

Edmond, Halcreek, and Willas all looked towards Davis. He had been kissed by a siren... His lips. They must have kept his lips after.

"I've never seen it before," the researcher said, "but I've read accounts of it. And from everything I have read, no one comes back from that. He belongs to the sea—to the beast that owns him now."

"We can save him." Halcreek stopped folding his arms, his hands balled into fists at his sides.

The researchers looked at each other and spoke softly in their native language.

"Hey!" Willas interrupted them. "Mind sharing with the group?"

The main researcher they had been speaking to spoke up, "He says you are better off just tossing him into the sea."

Halcreek took a step forward before he was stopped by Edmond. He growled, "Better off tossing *you* into the sea!"

"There must be a way we can help him," Edmond spoke over them.

The researcher shook his head again. "I am telling you: That man doesn't even hear you. All he hears is her calling him back to her. She is all he sees now. He will do anything to get back to her—even if it means killing you."

Edmond and Halcreek exchanged looks. The news was grave, but how true was it? They were only old accounts he had read. Maybe there was a chance Davis could be rehabilitated. They wouldn't be tossing their own man overboard.

Edmond looked to Davis. He writhed and jerked under his restraints. He cried out every so often, but of course, no one knew what he was saying. Was he really too far gone? Edmond didn't want to believe it.

"How do they do it?" Edmond found himself wondering aloud. "How do they bewitch men like this?" It just wasn't fair. Using their loved ones against them to the point they think they can reach out and hold them…

"That is what we mean to find out," the researcher told him. "How can we do that if we give her back to the sea? How can we learn anything without observation?"

Halcreek answered him, "They are creatures of the devil, and the devil is the king of lies and deception."

"Oh please." The researcher rolled his eyes. "Keep your gods and devils. This is science. They are flesh and bone just as we are. This is nothing more than their defense mechanism or hunting strategy. They need a way to protect themselves and a manner in which to feed."

"Flesh and bone that keeps us locked in purgatory?" Edmond reminded them of their predicament.

"I don't know how they control the fog," the researcher said, unsure, "or if they even control it at all. There isn't enough research to prove anything. But from what I have read, either they bring the fog, or the fog brings them."

"Your idiocy brings them," Halcreek sneered. "I say just toss all three overboard. They can get real up close and personal for their research then."

"We need a safe and clean strategy for moving her," Edmond declared. "There is no telling how much damage she can do if we come into contact with her."

"Well, we can't just move the box itself." Willas sighed. "We would need equipment for that."

"So we will have to open the box." Edmond didn't sound too happy about it, but he knew it needed to be done. "How was it that you captured her?" Edmond asked the men.

"Tranquilizer." The researcher shrugged.

"Tranquilizers that you don't happen to have any more of, I assume," Edmond sighed.

Both men shook their heads as an answer.

"Well, a net is out of the question." Willas took that option right off the table. "They can shred them too easily."

"Imagine how easily they will shred us if we try to lift her," Halcreek said.

Edmond had a thought and turned to the researchers. "Can…" He thought about how to ask this without looking foolish. "Can she be reasoned with?"

The researchers both looked confused. Had they never even tried speaking with the creature? "You mean… does she talk?"

Edmond nodded.

"We know they speak," the researcher told him. "They are insanely intelligent beings. But it would be like trying to speak to someone in Latin. You can speak, but that doesn't mean she will understand you."

"I've heard them speak English," Edmond reasoned.

"No," the researcher corrected him, "you heard what you wanted to hear. That is how their power works."

"We have to try," Edmond decided. "We can try to communicate. To let her know we mean her no harm and that we mean to return her to her family."

The five men entered the cargo hold and returned to the box. Edmond felt the hair on the back of his neck stand up. What if he was fooled again? He didn't want to feel that knife in his heart anymore.

The researchers removed the cloth from the box. One of them did a simple knock on the wood.

They heard a similar knock in return.

"Well," the researcher waved his hand toward the box, "say your piece."

Edmond stepped forward and placed his hand near the hole at the top of the box. "Can you hear me?"

"Daddy?" Emmalien's small voice called out from within.

Edmond took a step back. He couldn't deal with this. Not again. "You're not my daughter." He tried his best to focus and keep his mind clear. "I know you aren't."

"Please let me out," Emmalien pleaded softly. "It's so dark in here. I'm scared of the dark."

"We…" Edmond fought to keep his resolve. This wasn't his baby. "We want to help you. We want to let you out of the box. Return you to your family."

"Help me, Daddy," she continued her charade.

"You're not her," Edmond was talking more to himself than to her. He knew this wasn't his daughter, but why did it feel like it was? Why did this tug at his heart so strongly?

"Do you miss me?" she asked sadly, "Are you taking care of Mommy?"

Edmond felt tears welling up in his eyes. His throat became tight; his heart ached. He rested his hand on the box again. "Yes, baby," he answered, feeling heavy, "I'm taking care of Mommy. We both miss you very much."

"Captain." Halcreek approached him and rested a hand on his shoulder. "That's not Emmalien."

I know that already! He shook his head vigorously. He rubbed his eyes to rid them of their tears. He knew this wasn't Emmalien. He just wanted it so badly to be her.

"I miss Benji, too," Emmalien pouted from behind the wood.

"Benji?" Edmond looked up. Their dog. She remembered the large fluffy poodle she used to curl up with when they read her stories. How? How was any of this possible? Edmond looked at the researchers, "She… knows about our dog…"

"It's fascinating." The researcher looked utterly excited. "They can either read minds or memories, or they project our memories back to us somehow! Do they speak to you in English? They speak to me in my mother tongue! They knew the playground I used to play at as a child!"

"Yeah," Halcreek rolled his eyes. "I bet they are great kissers, too."

"This is a phenomenal concept." The researcher didn't take too kindly to Halcreek's quip.

Halcreek gave the box a solid slap. "Come on! You want us to help you or not?" After a few moments of silence, Halcreek responded with, "Look, sweetheart, that isn't going to work on me. I had a nasty divorce, and I don't want anything to do with women until the day I die. So, let's just focus on the topic at hand. We help you; you help us."

"You'll help me, Daddy?" Emmalien was back to speaking to her father.

"Yes, baby." Edmond simply couldn't help himself. He felt like he was going crazy. This had to be his daughter. Maybe a piece of Emmalien's spirit existed somehow within this siren. Maybe sirens had a way of being a conduit for the dead. Maybe he was really speaking to Emmalien in these moments. His actual baby girl was wrapped up in the scaly skin of a siren. It sounded crazy, but what if?

"I want to go home," Emmalien told him. "Can we go home now?"

"Yes!" Edmond rested his forehead on her box. He felt like he was talking through Emmalien's coffin again. Back when he begged her to come back to him back then, there was no response. But now… "Yes, we can go home. That's what we all want, baby. We all want to go home."

"You'll come with me?" Emmalien spoke just on the other side of the wood, so close he could feel her voice echo off the wood. "I'm scared to go alone."

Come with her? What did she mean? To join her in the ocean? Was that possible? Could he have a life with his daughter again?

"Hold my hand, Daddy…" Her small, delicate hand reached from the box again.

Before anyone could stop him, Edmond reached up to grasp her hand into his. It was warm and soft. She clutched his fingers.

The men around him jumped to his rescue, only to find that nothing happened. She was simply holding his hand.

"Let's go home, Daddy," she begged him softly. "Together."

Chapter 14

July 14, 1834

The researchers brought a siren aboard. That is where the crying came from. That is why we are hunted down. Stalked. They follow us every time we sail. Our speed is irrelevant. They find us.

Hammond was taken. Another one who jumped into the sea when they sang for him.

I have Maryann calling out to me from the waves and Emmalien calling out to me from cargo. One tempts me to jump while the other keeps me from going over the edge. I get torn between the two.

I can almost understand Davis in his madness now. I can only imagine what he hears when they call to him. I see why he fights to jump off this ship.

One of these nights, it will be me.

Edmond slept right next to the siren in her box that night. Though, to him, it wasn't a siren. It was a second chance with his daughter. He fell asleep while talking to her. She loved it when he would tell her stories from his adventures. She wanted so badly to sail with him, but Maryann wouldn't have it. She didn't even like Edmond sailing much these days. Now, he couldn't blame her. He made a promise within himself to stay with Maryann forever. He'd never sail again if she asked it of him.

When morning came, he ran his hand across the box. He wanted to touch her face and kiss her hair. He spoke to her. Was she awake? "Baby?"

"Are we going home, Daddy?" She was.

Edmond smiled. He was so happy to hear her voice. He sat up and peered into the box. Her little pale face came up to greet him from the other side of the hole. She was perfect, just like he remembered her.

He felt a dread rising in his gut. He was stalling with this plan to cast her over the side of the ship. He didn't want to lose her again. It wasn't Emmalien, but… it was.

The men would want answers on how to carry out the plan. He would be expected to give them answers. He couldn't stomach the idea of being separated from her.

What if…?

What if he somehow managed to take her home to his home—his real home?

What if he took her back to Maryann?

She would be scared at first, but if she just saw this for herself…

She would see that they could have their daughter back. That they *did* have their daughter back, in one form or another.

He groaned as Willas came through the door.

"Captain," he greeted him. "Thought I'd find you here."

Edmond didn't want to have this conversation. He wasn't even sure how to. He wanted to tell Willas his plan, but he knew he couldn't. Willas would protest and try to tell him this wasn't his daughter, but he already knew that. He wasn't stupid. He also knew that this was the closest thing to his daughter that he had.

Edmond let out a long yawn. "Tell the men to sail."

"Sail?" Willas sounded and looked confused. "I thought we would return her first?"

"Not yet." Edmond tried to sound nonchalant. "Maybe we can just… Maybe we can still make it home."

Willas saw right through Edmond. "We can't sail anywhere while she is on this ship, Captain. She isn't little Emmalien, no matter what she says. It's a delusion. You see and hear your daughter, while I see a monster and hear her hissing in your ear."

Edmond scoffed. Willas couldn't understand. He had no children. He couldn't understand the weight of this and how much it hurt Edmond. He couldn't understand how much he felt as though he himself would die from heartache when he lost her the first time and how scared he was to lose her yet again. He couldn't survive it a second time.

"What do you mean to do?" Willas tried to reason with him calmly. "Keep her moisturized in your bathtub at home? You really think Maryann is going to let you come home with… Well... a literal siren?"

"She has her memories!" Edmond insisted.

"No, she doesn't," Willas repeated what the researchers had told them previously, "It's a delusion. It isn't real. She is reflecting your own memories back at you. She uses your mind against you."

"No." Edmond shook his head. He couldn't hear that. "I wasn't thinking about Benji. I haven't thought of him this entire trip. She knew, Willas! She knew things only Emmalien could have known."

"Because of their spells." Willas didn't know what else to call it. "Their… magic. It is all in your head. Everything you see and hear is all a fantasy they fabricate within your mind. You see your daughter; I see a strange woman I have never met before. You hear her calling you. I hear her weeping. She plays different visions for all of us."

"I can't explain it to you." Edmond felt defeated. He couldn't explain it to himself either. All he knew was that somehow, Emmalien lived on through this creature.

"We have to give her back." Willas patted Edmond's back. "We can't keep her. You can't keep her. Emmalien is gone. You must let her go."

Edmond felt the pain in his chest again. The burning in his eyes. The knife in his windpipe. This time, he didn't blink the tears away. He couldn't. Instead, they came trailing down his face, one right after the other, until the tears came in sobs that rocked his shoulders.

He hadn't let himself cry in a long time, but today, he couldn't make the tears stop even if he tried. Willas rested a hand on the captain's shoulder for some amount of comfort.

"Why do you cry, Daddy?" Emmalien sounded as if she would cry soon.

Edmond's hand shook when he rested it next to the hole in the box. "I'm so sorry, baby. I wasn't there," he sobbed. "I should have been there!"

"You're here now." She reached out and rested her little hand over the back of his. "Please don't leave me again."

Willas cringed at the sight. He saw it for what it was in that moment. His captain, bewitched by a young siren. He wanted so badly to pull his captain away, but part of him felt like Edmond needed this.

"I won't leave you," Edmond cried into the box, looking down at her face, "I'll never leave you. I love you, Emma."

"Captain." Willas felt it had gone a little too far. He wanted Edmond to get closure, but he didn't want Edmond to get lost in his delusions. He had a hard time finding where the line was. How much of this was helpful? How much was harmful?

Willas gave Edmond's shoulder another pat. "Come on." He tried to encourage him to get up: "We need to find Halcreek and the researchers. We need to discuss how we'll return her home." He tried to speak delicately. He wasn't sure just how far-gone Edmond was. He just knew he had to be careful. His mind was being played with like putty by these things.

"I want to stay with her." Edmond wiped his tear-stained face with the back of his hand, trying to clean himself up.

"We need a plan, Captain, and you need food. Come on." Willas did his part to help Edmond up to his feet. He let Edmond compose himself as best he could. Then, both men walked out of the cargo hold. All the while, Edmond kept looking over his shoulder longingly.

Chapter 15

Edmond and Willas walked across the deck. Halcreek had been awake all night on patrol, as usual. There didn't seem to be any bloodshed from the sirens. They found Halcreek at the back of the ship.

"Get to shoot anything?" Willas asked him, then tried to peer at the water through the fog.

"Not this night." Halcreek almost sounded disappointed.

"Captain!" All three of them heard the call from somewhere on the deck.

"Here!" Edmond returned. "At the helm!"

After much shuffling and clanking, Piles emerged from the fog, walking with makeshift crutches. "Captain!" he greeted Edmond a little breathlessly. "Reporting for duty!"

Edmond couldn't help but smile. He closed the gap and gave Piles a strong hug. He was so happy to see him up and about again. Even the color had returned to Piles's skin. He looked and sounded so alive. It brought Edmond's heart joy.

Halcreek smiled as well and clasped Piles on the shoulder.

"What are you doing?" Edmond released the man.

"Trying to work." Piles readjusted the planks under his arms. "I swear, Captain, I am still useful! I can still work!"

That was when the realization hit Edmond along with a hollow sadness. Piles had known ships for as long as he could walk. His whole life had been spent working on a deck. It was all he knew, and now, with one leg and no free hands, what was Piles to do? He could barely keep his balance while the ship was idle. He could only imagine how unstable he would be when they started moving again. Piles's whole life had changed, Piles just hadn't realized it yet—still desperate to prove his worth.

Edmond couldn't stop the sad frown that appeared on his face as he thought of how to respond.

Piles was quick to convince his captain, "Honest, Captain! Any job! I can do it as well as any man!" Edmond's heart went out to him.

After some consideration, Edmond thought of a job that Piles could do while seated. "I do have one important job," Edmond told him, "but it isn't pretty."

"Anything," Piles assured him.

Edmond nodded. "I need someone to monitor Davis. It won't be easy. You'll watch a man under the spell of madness. He is strapped down for now, but he needs to be fed and cleaned up. His bedding needs changed."

"I can clean and feed him!" Piles agreed.

"No. Not you," Edmond corrected him. "It's too dangerous for one man alone. You'll only monitor him. I'll send someone down later to help with the rest."

"But I can do it, Captain." Piles sounded desperate. Of course he was. If he wasn't a functioning member of the crew, then what was he?

Edmond felt the need to reassure him, "I know you can. You're as capable as any. But for now, only monitor him."

Piles didn't look happy. He looked hurt, as if he were simply being cast aside to sit in a corner while the men did the real work. He nodded reluctantly, but made his way down, shuffling unsteadily along the way.

Edmond, Halcreek, Willas, and the researchers met in Edmond's quarters. It was time for a plan, but Edmond would much rather be sailing. How was he supposed to keep her? How could he convince Willas and Halcreek?

"The fog may follow us, but eventually, we will reach land." The researcher offered his plan first: "If we can just get her to land, we will all be safe."

"That's not the plan," Willas told them. "Our plan is to return her, not take her with us."

"That is your plan," the researcher disagreed. "I have my own plans."

"We have shackles in the brig," Halcreek reminded them. "I say we use them. If she is docile, maybe we can reason enough with her to use them without a struggle."

"And if she isn't docile?" Edmond really wanted an excuse to keep her.

"There are five of us and one of her," Halcreek pointed out. "If push comes to shove, we may overpower her. Plus, we have rifles."

"We can't kill her!" Edmond was quick to spit out.

"Only to threaten her." Halcreek shrugged. "They are smart enough to know what guns are. Let's use that to our advantage."

"Great," Willas chimed in. "We use shackles to help with the hands, and I say we also use sheets. Wrap her up and drag her to the lifeboat. We can lower her back down to her family safely. Returning her kindly can show them our regret."

Edmond tried to pick this one apart. "If she shreds the sheets?"

"I don't think she will." Willas thought out loud, "She wants to go home as badly as we want her to. Therefore, a little explanation can go a long way. She should let us if we communicate our side."

"It seems dangerous," Edmond mumbled. He wasn't sure of how to stop this without looking insane.

"This is our life's work we are talking about!" the researcher proclaimed. "You can't do this! I am telling you, sail until we reach land. Just sail!"

All of them looked to Edmond. The final call was his. On one hand, he faced ridicule from two researchers desperately trying to hold onto their evidence. On the other hand, he faced potential mutiny. His men looked to him for reason and protection. How would it appear for him to suddenly side with the researchers? To put all his men at further risk just for… for what?

He rubbed his temples, stalling. He had to think. How could he pull this off?

"It is possible…" he began with a hint of uncertainty, "that land isn't far off…"

Willas already knew this was coming. "In which way?" he asked without expecting an answer. "Which way do we blindly sail in hopes of not crashing into something? For how long, Captain?" He was becoming increasingly more agitated. "How many days will pass before you are ready to release her at last?"

Never. At least, that is what Edmond wanted to say. He knew better. Logically, he knew Willas and Halcreek were right, but his emotions were eating him up from the inside out. He wanted to scream into the void.

"What are we waiting for?" Halcreek asked.

"We need our captain to agree." Willas looked to Edmond.

Edmond couldn't bring himself to say anything. He only looked down at his desk.

"You can't do this, Captain Conway!" the researcher begged. "We will double the pay! Triple the pay!" He leaned on the captain's desk in desperation, not realizing he was fighting for what Edmond wanted already.

"I'll tell you where you can shove that triple pay!" Halcreek snapped. "You realize men are at risk? We are practically out of food! We will all starve!"

"Do you know how many lives we can save by keeping her? We can understand their powers, understand how they learn, how they grow, how they communicate." The researcher pointed out, "This will really change the world!"

"Not on my watch!" Halcreek shot down the whole idea of it.

"Tell them." Willas looked to Edmond. "Tell them we will return her."

Edmond felt like he was backed into a corner, torn between the two options. How much he wished he could just let down the sails now, but their food supply was so thin. They needed to act before the sirens or starvation did.

He felt a ringing in his ears. A pressure squeezed his head. They all looked to him for an answer. He didn't know what to say. The words just wouldn't come out. He couldn't physically tell them to get rid of her. It seemed so alien to him that, not so long ago, he was demanding she be thrown from the ship. He felt like that past him was a totally different person somehow. How could he have been so blind as to what these creatures were?

Screaming could be heard from outside the door. "Captain! Captain!"

DeVitas entered the door. He hadn't even bothered to knock. He was still screaming, "Davis!"

There were no questions asked. It was clear there wasn't time. Edmond, Willas, and Halcreek all rushed from the room toward the holding cell.

Inside, Keller was struggling on the ground with Davis. Blood and feces were smeared all over the floor and walls. Within the bloody smears on the floor lay Piles in a puddle of his own blood. His face and neck had been chewed to a point where he wasn't even recognizable anymore. He was still gasping for air.

Edmond and Halcreek rushed to his sides while Willas worked with Keller to strap Davis back down.

"Piles!" Halcreek screamed at him. Edmond grabbed Piles's hand just to find that it, too, had been chewed up. The man's face and neck looked like ground meat with teeth. Blood was still pooling around his head.

Piles sputtered, blood leaking as he spoke, "I thought…" He shivered, his words shaking as much as he was. "I thought I could do it…" The mess of meat around his mouth mushed as he spoke.

"You damn fool!" Halcreek screamed over him. "We told you to wait!"

Piles's eyes were red as he looked around the room without eyelids. Tears mixed with blood seeped into his butchered face. The blood from his neck pulsed out to join the rest on the floor. Every time he tried to swallow, the muscles in his neck could be seen moving from under the broken skin. He took a couple of ragged breaths just to gasp and go still.

Even Davis's screaming seemed tuned out while Edmond looked at Piles's disfigured remains. That's all he was now. Remains. So desperate to prove he was still capable that he got himself killed. Edmond should have known better than to put Piles over such a task. The puddle of blood was soaking into Edmond's pants.

Edmond slowly let go of Piles's hand and rested it on his chest. His eyes went down to Piles's leg, or rather, the stump that used to be a leg. All the work Halcreek had done to try to save the man's life... all the pain and suffering… for what?

Halcreek stood. His hands ran through his hair, and he let out a frustrated scream that shook the air around them. He swung his fist to punch the wall. He seemed feral. He stormed over to Davis, who was strapped down, and wrapped his burly hands around the man's throat to strangle him.

"No!" Edmond was up in a flash to grab Halcreek. Willas and Keller both tried to yank Halcreek's hands free of the man's throat.

DeVitas joined in to pull him off. It took all four of them to rip him away from Davis at last.

"I'll kill him!" Halcreek roared. Edmond did his best to stand in front of Halcreek to prevent him from going after Davis again.

In all the commotion of trying to calm Halcreek down, another sound was heard. Back in the cell, strapped to the cot, Davis cackled joyfully.

Chapter 16

"Throw him overboard!"

It had been a few hours since the attack. Although Halcreek had calmed down, it was only slightly.

"No," Edmond told him for the hundredth time. "He's mad. He's not a criminal."

"His mind isn't his own," the researcher reminded them. "He will kill any of you. It is best to toss him off the ship."

"I've heard enough of this!" Edmond slammed his fist onto his desk. "The next person who tells me to throw my own men overboard is going over themselves! I won't have it!"

"He killed Piles!" Halcreek yelled.

"Piles didn't listen!" Edmond hated to defend Davis's actions. He knew exactly why Piles had done it. That drive to feel useful and independent got him killed. It would be necessary to give Piles to the sea, although Edmond didn't feel right handing one of his men over to those things. What if they ate him? It just didn't feel right. It would be like tossing Piles to the sharks. He was too good of a man. He deserved better. His body was wrapped in sheets and lying in the sick bay for now. He had just gotten up, only to be there again.

"We should sail," the researcher interjected. "Let us leave this place and go to land."

"What about tossing him over?" Halcreek suggested. "He isn't one of us."

"Easy." Willas tried to calm the situation, so Edmond didn't have to. "I think our first manner of business is—"

Edmond interrupted him, "We sail."

"Captain!" Willas sighed.

"It's my ship!" Edmond declared. "My men, my decision! We sail. Halcreek, let down the sails."

Halcreek hesitated, looking to Willas. Willas had nothing to say. He only looked away. Halcreek gave a reluctant nod. "As you wish, Captain."

Emmalien's Constellation was on the move again just as night was beginning to fall on them. Edmond hoped beyond hope that they would reach speeds that couldn't be followed by the sirens. Perhaps if they carried on without fail, without a single pause, they could outrun them.

He was beyond stressed. He needed a break from it all. He needed Emmalien.

Edmond went to the cargo hold and sat on the floor next to her box. He could hear the water sloshing around within. He hoped she was comfortable and not too cramped.

"Baby?" He leaned against the box, hoping to hear her voice.

"Daddy?" she said softly.

His heart fluttered. "We are going home now, baby." He told her. "We are going back to see Mommy and Benji."

"I want to go home…" she whispered.

"Yes. I know. We are." He nodded. "There's something else." He said, "Mommy will have a new brother or sister waiting for you." He smiled at the thought. His heart ached that he couldn't be there for the birth, but he was looking forward to taking his little one into his arms at last—the same way he had taken Emmalien into his arms so long ago.

Emmalien was quiet.

Edmond frowned. "Emmalien?"

"Daddy." She reached her tiny hand out from the hole. "I want to go back home. Back to the water."

Edmond took her hand. His heart hurt even more now. Emmalien was confused. She was scared. She was stuck in the wrong body. Once she was home, she would be happy again. She would love to see her mother again. He knew it.

"No, baby." His thumb stroked her pink hand. "I am talking about *our* home. Back at the harbor."

"I can't go to the harbor." She sounded like she was about to cry. "They'll kill me. The men will want me dead."

"No!" He shushed her. "No, I won't let them see you! You'll be safe! I'll keep you safe, little one!"

"You can't take me there," her voice broke. "Daddy, I'll die."

"You won't die!" he was quick to say. "Don't say that! Please!"

"Daddy…" Her hand grew cold and webbed between his fingers. He opened his hand to see the small, gray, scaly hand. He investigated the hole in the box, only to see the bulbous eyes peering at him from the dark. "Look at me, Daddy." She still had Emmalien's voice. "I must go back to the sea. I'll die."

With each splash over her face, her color returned to pink, and Emmalien's bright green eyes peered back at him. Her hand was soft, pink, and warm again.

Edmond felt like a knife was stuck through his chest. His stomach was in tight knots. "But…" he croaked, "I can't lose you again."

"Come with me," she cooed. "Come with me into the sea. We can be a family again."

"Maryann…" He shook his head.

"Mommy waits for you. Come and see." She smiled up at him.

July 15, 1834

Davis killed Piles. It was a bloody mess. Halcreek is furious. I can't blame him. Not sure what to do with Davis. For now, he remains restrained in the holding cell.

We are moving again. Hopefully, we will continue to move until we have reached land.

Emmalien wants to return to the sea. She calls that her home. I blame myself for always telling her tales of the ocean. Maybe this was my curse or my blessing: to lose my daughter just to find her again at sea. She wants me to go with her. To join her in the ocean. She says Maryann is there, too. Could they really be channeling our loved ones? It feels so real.

How I wish I were home already.

Edmond spent another night next to Emmalien. Except this time, he was joined by the researchers who asked him various questions about what she was saying to him. Evidently, when a siren spoke to people, only one person could hear it. They were speaking to people individually, the same as when they sang for specific people.

"You are doing the right thing," the researcher told him. "We should reach land eventually."

He had half a mind to tell him that he wasn't doing it for him. Although, he was in another dilemma. How would he even be able to keep her for himself if these men would want her back?

He looked down at her face in the hole. She was begging to be released. Was that the right thing to do? Should he just let her go to the ocean? He knew she had family there, but she was also his own family. He was torn yet again. Even his own daughter sided with his men.

"I think…" He rubbed his eyes tiredly. "I think we have to give her back."

"What?" The researcher perked up now from his notebook. "But I thought we would sail?"

Sailing meant that he would eventually have to fight these men for her—unless he just tossed them over. He could say they lost them in the storm.

But that wasn't Edmond. He wasn't a murderer.

Sailing also meant that Emmalien could potentially die. She had already been stuck inside this cramped box for so long. She needed fresh water. She needed food. She needed room to move freely. She wasn't in a safe situation.

"I don't know." Edmond was still wrestling with himself. "She wants to go home."

"Of course she does," the researcher scoffed. "I am sure most animals in the zoo want to go home, too, but this is necessary."

Edmond didn't know how he would take her home with him, but he figured the ocean would be better than whatever these men planned to do with her. Running their tests and experiments… She deserved a better life than that.

Edmond groaned. He would have to go back on his word. He would have to stop this ship and release her.

At the same time, he was so terrified for her to disappear under the water, never to be seen again. What if she just left him? What if all this really was some trick from a siren to get released? Did she even really want him to join her?

"I need sleep," the captain told them. "Please. Leave me with her tonight. We will talk more in the morning."

The researchers were more than hesitant to leave, but they did as the captain wished. He lay on his back next to the box and let all his overthinking take over. He would sleep on it tonight, then decide what to do in the morning.

"Baby?" He tapped the box. She tapped back. "I love you. You know that?"

"Yes," she spoke just above a whisper. "I love you, too."

Chapter 17

When morning came, Edmond was quick to find Willas and Halcreek. Both of them had been on patrol the night before.

"We need to give her back to the ocean." He finally said the words that Willas was waiting to hear. Willas smiled and gave the captain a pat on the back.

"Finally, you see sense," Halcreek teased him. "What changed your mind?"

"Emmalien," Edmond said at first, before needing to correct himself in front of his men: "The siren asked to go back."

"You realize you have spent your nights talking to a fish?" Halcreek looked worried.

Willas nudged Halcreek hard with his elbow, then returned his attention to Edmond. "We are happy you finally agree with us. Raise the sails?"

"Raise the sails." Edmond didn't like the sound of the words coming out of his mouth. Every second was another second closer to him losing his daughter yet again.

"Collect what you need," Edmond told them. "I'll… say good-bye… to the fish." He gave them a grim smile, then headed back down to her.

"Good morning, baby." He knocked on her box. "Are you ready to go back home now?"

"Oh, Daddy! You mean it?" She splashed around in her murky water. She was beyond thrilled.

"Yes," he said sadly. "You'll go back to the ocean today." He frowned as he thought of what to ask her next. "Will you be able to release us? Release this ship from the fog? We want to go home, too."

"I can't make promises." She peered out at him. "Mommy is very angry."

He thought of Maryann. She was probably angry with him, too. He was taking far too long to get back to her.

"You will be returned safely. Isn't that enough?" he tried to reason with the child.

"I don't know."

His men came through the door with blankets and shackles. Edmond hated the rattle of the chains. It seemed so unethical.

"Baby." He leaned close to the box now. "We are going to let you out. Can't you at least promise me that you won't try to hurt anyone? We want to help you. Just let us take you out. Please."

"I'll be good, Daddy. I just want to go home." The sound of joy and hope in her voice almost broke Edmond's heart. It was like she was happy to be away from him or something.

Halcreek and Edmond used crowbars to crack open the lid. They pried with all their might, and the wood gave. The lid with the hole popped clean off. There she sat, exposed for all to see.

She was a miniature version of the monster Edmond had seen only a few nights ago. She wasn't using her illusions at this moment. Her tail stretched up and out of the water. It was pale gray and looked razor sharp as it reflected the light of their lanterns. Her hair was inky black and draped over her shoulders. She was dripping as she sat up in the box, finally able to stretch her arms and webbed fingers out.

The sight took Edmond by surprise. This thing… *This* was what he thought was his daughter? Or… this *was* his daughter? Just in another form?

Halcreek held the shackles up. "We need those arms."

"Wait." Edmond stopped him. "She said she would work with us."

"Oh, I forgot we have the siren whisperer." He rolled his eyes, clearly annoyed.

"Work with us how?" Willas asked. "She can understand us?"

Emmalien looked to Willas. There was no telling what she was saying to him.

"I guess she does understand us, then," was all Willas said.

"Fabulous." Halcreek pointed to the blanket. "Then if you wouldn't mind, please come onto the blanket yourself." He didn't think it was a good idea for any of them to come into direct contact with her flesh. Even though Edmond hadn't any complications from holding her hand, there was no telling how their abilities actually worked.

She obeyed. She seemed happy to. She crawled her way out of the box and onto the blanket. She wrapped her long tail around herself and folded her arms around where her knees would be.

"We are going to lift you," Halcreek told her.

Together, the men walked with the blanket between them. With all three of them, she barely weighed a thing. The ship had come to a decent stop by now. They took her to the lifeboat and gently sat her down within.

"We're going to lower you down," Edmond told her. He was starting to feel the sharpness in his throat again. He was doing his best not to cry. They had to hurry before the researchers woke up.

In the blink of an eye, Emmalien appeared herself again in front of him, as pristine as the day he had left her. "Will you come with me, Daddy?" She smiled so sweetly.

He was speechless. He wanted so badly to tell her "yes" and climb into the boat with her.

"Let's go." The men began to lower her. Edmond didn't speak. He couldn't. The pain was too much. If he dared to say a word, he knew he would break down.

She sank past the edge of the ship and gradually came closer and closer to the waves. Even when the boat was on the water, she didn't move. She still cast her eyes up at Edmond. He never took his eyes off of her. His baby girl… They stared at each other for a moment. Finally, she turned and leapt into the water, disappearing beneath the rippling waves.

Edmond's heart sank. He did what needed to be done for himself and for his men, so why did he feel worse than before? Had he made a terrible mistake?

"Captain!" The researchers both ran to them breathlessly. "What are you doing?"

"It's done," Edmond said solemnly. "She's home. Keep your salary. We just want to go home."

"How could you do this?" The first researcher looked over the side of the ship. The life boat still bobbed about against the side of the ship. The man ran his fingers through his hair, threatening to rip it out. "Why?"

"She's better off." Edmond felt defensive over her, still feeling that fatherly cloak over his shoulders. He knew he had saved her from death. "This is my ship, and my decision was final."

"No!" the second man screamed, looking over the edge. He slammed his fists on the side of the ship before storming his way over to Edmond in a fury.

He didn't get a single word out before Halcreek inserted himself in front of his captain and shoved the man hard. The researcher slammed into the side of the ship, very near going over.

"Watch it." Halcreek made it clear that the shove was just a warning.

"You've thrown away everything!" the first researcher yelled back to Edmond while helping his colleague up.

Didn't Edmond know this already? She was his daughter—his everything. His heart was still an open wound because he released her. "Make yourselves comfortable," Edmond told them. "We'll set sail when this fog clears."

"Will it?" Willas asked, confused.

"We will see."

The researchers both stormed off below deck. Edmond was very much hoping he wouldn't have to see those men again until they reached land. It was a shame that they lost the salary, but hopefully Maryann would be understanding. That is, if she would even believe everything he had seen.

Night was falling, yet the fog still remained. Edmond stared off into the fog, half hoping he would see his little Emmalien again. Where had she gone? Surely, she was still near if the fog remained. At the same time, why? They had returned her to the ocean. Shouldn't they be free to go now?

"It doesn't look like it is letting up." Willas sighed. "Maybe... maybe it was all a lie?"

Edmond frowned. He couldn't think that. He had had a real conversation with her. He could feel that. She did mention that her mother was angry. His Maryann was a fierce mother, too. He understood how protective she was. Maybe that was what this siren was channeling. Was it her mother who controlled the fog, then?

"Should we sail anyway?" Willas turned to face Edmond.

"No." Edmond wanted to wait to see if the fog would clear, but at the same time, he hoped he would see her again. Once more, he felt torn between his men and his daughter.

Willas chuckled. "Can you believe this was all for nothing? No way we are seeing a single coin from those men."

"Good riddance," Edmond scoffed. "If I never have to see them again, it would be too soon." He suddenly stopped when something in the water caught his eye. Something was below the ship. He saw the length and fins of a tail. It trailed near the ship, just to dip back into the black water.

"What?" Willas looked below, too.

"I saw… something." He wasn't sure. A siren, or just a fish?

He almost let it go entirely, until he noticed another.

And then another.

"Willas," Edmond nodded toward the water.

They both saw them. Dozens. Heads barely peeking out from the water, their eyes reflecting up at the men. They were surrounded yet again, which Edmond figured would happen once they had stopped. He hoped that the sirens would understand their actions today, parent to parent.

"Captain!" Keller called out. "We're surrounded!"

Edmond heard it again. The singing. He turned back to the water to see Maryann and Emmalien both peering at him from the dark. She had returned! Maryann brought an infant bundled in a damp blanket up to her face.

His family.

His entire family, just below.

They were waiting for him. Just like Emmalien told him they would.

"No!" Willas threw his arms around Edmond and yanked him hard to where they both fell backwards onto the deck. He hadn't even realized that he had started climbing the side of the ship to jump. "Halcreek!" Willas called out for help.

Halcreek struggled to keep his rifle in hand while he wrestled with Edmond. They both fought to drag Edmond away from the deck and below to his own quarters.

"That's my family!" Edmond was screaming at the men. He had to hold his baby. He just had to.

Screams were heard from behind Halcreek and Willas as they guarded the captain's door.

"What now?" Halcreek groaned.

Chapter 18

Keller came to find them at the door, blocking in Edmond. "It's the researchers!" He gasped for breath. "They released Davis! He's going for the edge!"

Edmond snapped out of his desperation. *Not again. Not Davis.* Those damned researchers and their petty idiocy!

The men rushed forward to find the madman. He would either get himself killed or be killing another. Edmond could see Piles in his mind again. The poor man deserved better. He couldn't have more blood on his hands. Davis just needed help! He was as much a victim of these circumstances as the rest of them were.

They finally came back to the deck to find DeVitas trying his best to keep Davis from jumping over the edge of the ship. He struggled to keep a grip on him, yanking at his clothes and arms. Finally, he pulled him closer to shout into his ear.

"Snap out of it!" DeVitas yelled, bear-hugging Davis.

A big mistake. That was far too close of contact.

With a wild roar, Davis leaned forward and took a sickening bite out of DeVitas's throat. The muscle and tissue gave an audible crunch as he came away with a chunk of flesh between his exposed teeth. The blood soaked Davis's shirt further. He held onto DeVitas's shoulders while he viciously went back for more.

"No!" Halcreek dropped the rifle and ran towards them.

Like some kind of rabid animal, Davis chewed away at DeVitas's neck. DeVitas couldn't even scream. He didn't have vocal cords anymore. It was as if he was in total shock, frozen in place while he was actively being murdered.

Willas ran at them, too.

Everything was happening in slow motion again.

It seemed like DeVitas took a lifetime to drop to the ground once Davis was satiated.

Halcreek and Willas barely moved at all while running full speed to Davis.

The ship seemed eerily still on the waves.

Even the wind seemed to stop altogether.

Davis threw his head back and screeched at the sky before lowering his head to face the men. It was as if he had eyes again, somehow seeing them and glaring at them all.

With a vision no one could understand, he barreled wildly to meet Halcreek, teeth open for an attack.

A loud gunshot cracked the still air, ripping through the slow time barrier and bringing everything back up to speed. A spot formed on his forehead just as the back of Davis's head exploded outwardly; his head being thrown back again by the impact.

Halcreek suddenly stopped, in shock.

Davis hit the deck like a limp bag of bricks. His skull fragments and brain matter littered the deck behind him. A pool of blood began to collect under him.

Halcreek and Willas looked back behind them.

Edmond slowly lowered the rifle.

His hands shook so violently, it was a miracle he even managed the clean shot.

"Captain?" Willas looked at him while standing over the very dead Davis. Without his lips, he just appeared to be smiling extra wide.

Halcreek took this moment to run to DeVitas. The man was holding his own throat in both hands, blood seeping uncontrollably between his shaking fingers. He tried to speak, but only blood spurted from his lips. The man was clearly dying. There was nothing anyone could do at this point.

The researchers were on the deck now, looking flustered.

Edmond found his strength now to raise the rifle at them. They both held their hands up defensively.

"What have you done?" Edmond screamed at them.

"You released ours," the second man said with a shaky accent, "we released yours!" It was very bold of him to be making such a statement with a rifle pointed at him. They were so petty, they were willing to let others lose their lives. It was sickening.

"Shoot them!" Halcreek demanded justice as DeVitas was taking his last desperate breaths in his arms.

Oh, how Edmond wanted to. So badly. He wanted to put a bullet in each of their heads. They were useless to them now, anyway. How could they be so vile as to release a madman on them all? Was their research really worth that much to them? Edmond really had half a mind to end them right where they stood.

A loud screech could be heard over the water. This was the second time they had heard a siren scream, and it was just as chilling the second time around, too.

The researcher pointed to Davis. "You killed her pet." Just what they needed. It wasn't like the sirens didn't have enough reason from the kidnapping to want them dead. Now, the man they had made their plaything lay in a bloody heap on the deck with his brain scattered in pieces across the wood.

Edmond hadn't wanted to kill him. That was the last thing he wanted. Yet, he now blamed himself for DeVitas's death as well. His men were all dying, and each one was his fault in one way or another. He should have left while he had the chance. He should have sailed. Was it too late now? Were they all doomed?

As if the fog wasn't bad enough, it was suddenly starting to get thicker.

The ship creaked and groaned in the silent air, rocking on the water.

It was near impossible to see anything. Edmond couldn't make out anyone through the fog, even those standing next to him.

The wet slapping and clawing were heard again.

Edmond knew that sound.

They were climbing onto the ship.

"They're coming!" he yelled out to warn his men.

He could hear shuffling and voices around him, but he couldn't see a thing. Who it was for certain, he didn't know. Were they already on the ship? Was this an attack?

He moved through the fog in the direction that he last saw Willas. He wasn't certain how far off he was, but he managed to make it to the edge without finding him.

"Captain!" He heard Willas call for him.

"Here!" he answered. "Willas!"

After much stumbling around, he and Willas bumped into each other at last.

"Come on!" Willas grabbed the captain's arm. "We can't let them get to you!" He was yanking his sleeve to head below deck.

"I can fight!" Edmond shook his hand free. "These are my men!"

"Your mind is altered, Captain!" Willas pulled him again. "Please! We can't lose you!"

Edmond knew his mind was weak at times. He was torn between wanting to defend his men and to keep himself from jumping. While he wanted to fight for them, he knew he couldn't fight against their delusions. He followed Willas.

Willas hurried him into his captain's quarters. He shoved Edmond in quickly and locked the door behind them. "Sit!" He ushered Edmond to his chair. "I won't let anything happen to you." He took the rifle from Edmond as a means to defend him.

Edmond was appreciative of how much Willas cared for him. He knew that he only did the best he could to help out. Although, Edmond didn't feel right leaving his men above deck alone. They had one of the only rifles. They may have needed it up there.

What was he supposed to do? Hide like a coward? It didn't sit well with him. It just felt wrong. He tried to listen to any commotion he heard above them, but he didn't hear anything. He could make himself useful. He could at least document the situation.

He would update his logbook, just in case this was his last entry.

He sat at his desk and opened his book. Before he could even put the quill to paper, he heard Willas call out to him.

"Captain!"

Edmond paused.

That was definitely Willas who was calling out to him. He was sure of it.

Except something was wrong. His voice was muffled.

It was coming from the other side of the locked door.

Edmond felt a hand on his shoulder. He slowly looked over his shoulder to see Maryann looking down at him.

"My love," she spoke tenderly.

Chapter 19

Edmond froze in his chair.

"Open the door!" Willas yelled, banging hard on the wooden door.

"I need you." Her thumb rubbed Edmond's shoulder.

His heart began to ache. He needed her, too. He slowly came to rest his face upon her hand over his shoulder. He just wanted to feel her. He knew this wasn't her. He knew his sweet and tender Maryann.

"You saved our daughter," she spoke softly, her hand now holding his face gently.

"Always," he told her, his hand resting over hers. It was only now he noticed her other hand still held the rifle Willas had taken from him. Where was Willas?

The door rattled hard under the impact of Willas's shoulder, threatening to burst open with each slam.

"But there is still blood to be had," she told him. "One by one, I'll have you all until I get the right ones."

"The right ones?" Edmond looked confused. What was his wife talking about?

Suddenly, the door cracked, and splinters flew inward as the door flew open. Willas almost fell but managed to catch himself.

"Captain!" he screamed in a panic before raising the rifle he had.

The shot was deafening as it pierced the window next to Maryann's head. The illusion was broken. She screeched in fury as Edmond fell sideways from his chair.

With a strength Edmond didn't even know they had, the creature took the rifle she was holding and snapped it in half.

Willas readied his aim again.

The creature used her mighty tail to launch herself backwards, shattering the window. She flung herself back into the sea below, the glass singing as it rained down into the water with her.

"Captain!" Willas dropped to his knees next to Edmond to check him over. "Are you okay?" He checked his face.

"She didn't hurt me," he stammered. "Maryann would never hurt me."

"Come on." Willas stood and tried to help Edmond up. "We need to help the men." That sounded like the real Willas.

Together, they went back above to the deck. It seemed calm. There was no screaming, no panicking. Edmond needed a report. Once he found Halcreek, the news wasn't good.

"They took Shaw and Keller," Halcreek sighed. "Which means they took off with two of our rifles. That leaves us with only two."

"One," Willas corrected him, holding up his own. "The other was snapped."

"This attack targeted our defenses…" Halcreek realized aloud. They knew to go after the men with the guns. They were left open for an attack now, and it could come at any moment.

"And our captain," Willas pointed out. He looked to Edmond now. "What did she say to you?"

Edmond had to think about it hard to remember. It was like trying to remember a dream after you wake up—fuzzy and distant, only a word here and there coming back. "Something about the right one?" He looked confused.

"The right one?" Halcreek seconded his confusion. "Is that supposed to be you?"

"I don't know." Edmond shrugged. "'One by one,' she said, until she gets the right one."

"There's no one left to get!" Halcreek threw his hands up around them, indicating the run-down numbers.

"Let's get below deck and lock up for the night," Willas suggested. "Safety for now."

The next morning, the men didn't have to be told twice to let the sails down. Edmond was sure this time. He would miss his family, but he had to get back to his… family. It started to confuse him when he thought about it for too long.

Maybe if they could sail, they wouldn't have to lose anyone else.

However, the fog never lifted. Even though they sailed, they didn't seem to be getting anywhere. Were they doomed to be stuck until the sirens released them?

The more Edmond thought of it, the angrier he became, but not at the sirens. No. They wanted the right ones, and the only ones he could think of that fit the bill were the researchers. They were the ones who had kidnapped Emmalien to begin with. They were the ones that she must have wanted. It just made sense.

But what would giving them up mean? Tossing them over? Could Edmond really sacrifice men in order to escape this fog? What would happen to them? Surely, he would be giving them up to their deaths. Was he capable of that?

He left the helm to find Willas. He would have to tell him his thoughts. Through the fog, he came across Halcreek first, and then Willas.

That was good enough for Edmond. Now he could tell them both together. "I think…" Edmond wasn't sure of how to say it. If he was wrong, that meant he was carelessly throwing away the lives of two men he barely knew. "I think I know how we can get out of here."

"I'm all ears." Halcreek was more than willing to hear him out.

"The researchers," Edmond sighed, "I think they're the ones she is after."

"So we toss them over. Like I have been saying this whole time!" Halcreek snorted.

"It will be a fight," Willas told them. "They won't go willingly."

Edmond suddenly felt a sort of regret for even mentioning it. Was he really plotting to murder two men? Had things truly come down to this? While Halcreek and Willas seemed on board with the plan, Edmond was having his own doubts.

"It would be murder," he reminded them. Surely, he couldn't be the only one housing doubts within himself.

"It would be justice," Halcreek responded.

"And freedom," Willas added. That part wasn't lost on Edmond. To be free again… It meant returning to Maryann at last.

"I say we drag them up here kicking and screaming," Halcreek suggested. "Throw them over like sacks of screaming potatoes."

"Maybe we can reason with them." Edmond knew it was a long shot.

"Right," Willas scoffed. "Surely, they'll understand and go to the sea willingly."

Edmond sighed. "Pardon me if I wrestle with this idea of murder."

"They aren't innocent victims," Halcreek reiterated. "They killed DeVitas and Davis. They killed every single man that was taken by those devils. They brought those things to us! They would have gladly seen us all die in the pursuit of their research."

"I know." Edmond nodded. "I know it." Yet still, he was a man of God. He only hoped there would be understanding and forgiveness for this to come.

"Alright, then." Willas readied himself. "Let's go hunting for men."

Chapter 20

After a thorough and complete search of the ship, it was easy to determine that the researchers were tucked away in their room. The door was locked and, from the sounds of it, likely blocked as well.

"Open this door!" Halcreek rammed his shoulder into the wood. The door barely rattled. It was definitely blocked off. The researchers must have heard or suspected something to have barricaded themselves inside. Although, it wouldn't have taken a genius to understand that the odds weren't exactly in their favor.

Their voices could be heard. They were agitated and speaking to each other in their own language. Back and forth, they fought frantically.

The men outside their room still had the rifle, but what good was that to this blocked-off door?

"Just open the door!" Edmond tried. "We just want to talk." At least, that was what Edmond wanted. He couldn't control what Halcreek or Willas would do once that door was opened.

"We aren't fools!" one of the men screamed. "You mean to kill us!"

"Look," Halcreek growled, "you can't hold up in there forever. You'll only starve until you eat each other." From the tone of his voice, it almost sounded like he would have been amused to see that outcome.

Did the men have to be handed over alive? Edmond imagined that was the case. After all, if someone had taken his daughter, he would want them alive. He would show them no mercy, and these things were more monstrous than he was, he was sure.

Edmond wasn't sure which fate was worse: starving or being taken by the sirens. Would they be drowned? Was that fate too kind for men such as this? Did the punishment even fit the crime?

Edmond couldn't be the judge of this. At this point, it was basically kill or be killed. He had to make the decision to give these men up if he ever wanted to go home and if he didn't want the sirens to haul away any more of his men.

"We don't have that kind of time," Edmond pointed out quietly. Who knew when the next attack wave would come? They would take off with two more men regardless if they were the correct men or not. He couldn't risk that. He had to get this door open.

"We can do this reasonably, or we can do this like savages." Edmond tried to sound reasonable, but how could you reason men to willingly come to their own deaths?

It was no secret what their intentions were. They would need to use force to open the door.

"I'll take this door right off the hinges!" Halcreek yelled through to them in a threat before ramming his shoulder hard into the creaking wood again.

A loud gunshot was heard. The door splintered outwardly as the shot ran through the door and into Halcreek's arm, cutting out the other side cleanly. Blood splattered back along the wall from the impact. Soon, the crimson fluid dripped down his arm.

Halcreek gripped his arm, blood oozing between his fingers as he backed away from the door.

The researchers must have brought a pistol on board with them. The hole in the door wasn't a large one.

Halcreek spit curses at the men through clenched teeth. All three men were away from the door now just in case they fired again.

"Stay away!" Both men gave them various warnings as shuffling could be heard beyond the door.

"I'll kill them myself!" Halcreek roared, storming back to the door, his adrenaline pumping. Edmond and Willas pulled him back before he could catch another bullet.

Halcreek ripped the sleeve off his shirt in order to tie it over his wound. He winced at the pressure.

"We need a plan," Willas stated the obvious. "One that doesn't involve us getting shot at."

"I'm ready to burn them out!" Halcreek went to the extreme. "Light a fire. They can burn or come out."

"We aren't burning my ship!" Edmond yelled. "Be reasonable!"

"A happy middle;" Willas tried to balance the idea, "we can always smoke them out. Make them think there is a fire. Use a flame we can control."

"That won't be enough." Halcreek winced from his injury. "It needs to be convincing."

"They will still come out shooting," Edmond made the point. "We need to get that gun away from them."

"How do you propose we do that?" Halcreek rolled his eyes. His agitation was clearly growing as his arm pulsed from the pain.

Edmond didn't have the answer. At the same time, he couldn't stand the idea of losing one of them. Could they really just wait for them to come out on their own?

Even more so, could he possibly reason with the sirens? Could he tell them where to find the researchers to the point that the sirens would leave the rest of his men alone? Could they understand him like his Maryann and Emmalien did? He didn't know what to expect.

"Can't we just shoot them?" Halcreek was still very much bitter from his wound, his sleeve drenched with a dark red stain.

"No." Edmond could feel his pain, but they had a duty to their men. They had to finish this properly, and that meant handing them over in one piece.

"We don't know how much ammunition that they have," Edmond said, "or if they only have the one gun." He paced as he tried to think. So much was riding on his shoulders.

"They can only reload so fast." Halcreek leaned against the wall, his blood smearing from his shirt onto the wood. He cupped his hand over the wound again to apply more pressure. "I need to dress this." He walked away from them to go to the sick bay area. His supplies would come in handy.

Willas and Edmond went into the captain's quarters. He needed to update his logbook, and they needed time to think of a plan. It was highly unlikely the researchers would leave their room for any reason—even food wasn't worth the risk.

"I hate to admit it, but I am starting to side with Halcreek." Willas leaned against the desk. "I'm more of a 'think before you act' kind of guy, but maybe this is something we really should use brute force with."

Edmond could see that Willas was scared. He was, too. For Willas to side with Halcreek, he could tell the situation was getting out of hand. He had two rats on his ship and no way to properly collect them.

"That could involve one of us getting shot. Again," Edmond reminded him.

"The window," Willas said.

Edmond thought about that. There was an outer window that led to their room. It was possible they could use the window as a means of accessing the room, unless the men had boarded that up as well. Though, it wasn't likely. Maybe they could lower down to the window to at least get a visual of what was happening within the room. It was a long shot, but it could work.

"It's not impossible," Edmond pondered aloud.

"No, the window!" Willas pointed behind the captain. Edmond turned around to see fog pouring in through the broken glass. The fog was getting thick and concentrated again. Another attack? So soon? With no defenses? They would come to take more men—any man they could grab. Perhaps one that was alone and wounded.

Willas and Edmond both looked to each other at the same time before saying in unison, "Halcreek."

Chapter 21

The two men rushed as fast as they could down the hall to check on Halcreek. By the time they made it to the sick bay, they only had seconds to spare. One of the creatures was on top of him. They clawed viciously at his arms that he had raised to defend himself.

Willas wasted no time firing off a shot from the rifle. The round tore through the shoulder of the beast. It howled madly at the blow, turning its attention to Willas.

"Wait!" Edmond tried to pause the whole mad scene. "Stop! Just stop!" He was hoping he could take the time to speak to the creature, to reason with it, but after that rifle went off, it was too late for talking.

The creature used its tail to slither in on Willas. Edmond didn't know if the creature was going after Willas himself or their final weapon of defense. It didn't matter either way, since both options were terrible.

"Stop!" Edmond yelled again. "We can help you!"

Willas, by reflex, fired another shot at the approaching danger. The creature was too close to dodge. Half of the creature's head blew apart as it dropped dead at Willas's feet, its tail twitching for a moment as death embraced it. Its inky black blood pooled beneath it on the wood.

"No!" Edmond screamed out in horror. What had Willas done? He knew the man was only trying to defend his own life, but this? This was going to be war if the sirens had their say about it. There was no going back after this.

"It tried to kill me!" Halcreek sat up. The backs of his arms were a bloody mess. His shirt had been ripped in the struggle as well, with a few deep scratches blooming on his chest. It was clear that he was losing a lot of blood. "It got me good."

Willas rushed to Halcreek first. Edmond was still frozen, looking at the mess of a creature lying dead on the floor. Had that been his Maryann? Surely not.

"Tell me what to do." Willas sat the rifle down to help Halcreek dress his serious wounds.

Edmond came next to them to collect the rifle. He wasn't sure what to do with it now. Was it best to cast it aside to show them that no further blood would be shed, or was it too late for that? Edmond figured he knew the answer to that already. These things would want more blood for their loss. Before, it was a simple exchange for the kidnapping. Now, it was going to be hell.

"Easy!" Halcreek winced as Willas worked. "I'm not dead yet, I can still feel everything!"

"You're not going to die!" Willas demanded.

A screech could be heard somewhere above them.

They knew of the loss.

Screeching roared from every direction around them. They knew the water was alive with angry beasts. The whole ship would be swarmed at this point. They didn't have the defenses to hold them off.

"Barracade the door!" Edmond yelled to Willas. Both the men rushed to flip a table and push the heavy wooden furniture in front of the door. Nobody knew how long that would last, but it was better than nothing. Right now, all they had was a rifle and a hope that this wouldn't end in all of them being torn to shreds.

Screeching and claws splintering through wood could be heard all above and around them. They were definitely on the ship. Dozens of them were all over the place.

Willas and Edmond sat on the floor with their backs against the table to add more resistance while Halcreek did his best to work on his wounds. They were lucky there were no windows they needed to worry about. Edmond clutched the rifle, not even sure if he would be able to use it.

The researchers had had the right idea to barricade themselves in before. Maybe they would outlast this attack as well.

They could hear the creatures rampaging the ship. Breaking glass, splintering wood, flipping furniture. There was fury on the other side of the door.

Suddenly, there was a loud and forceful slam against the door. The door jolted open a couple of inches, sending the table forward into their backs. The two men rushed to push the table back into place and hold fast against the force of the next blow.

Each thrust pushed the men back a few inches, causing them to push right back to secure the hold. It wasn't clear how much longer they could keep this up.

"Help!" Willas called out to Halcreek. "We need you!"

Halcreek dropped his supplies and pressed his bloodied hands against the door. He stood above the men, pushing with all his might. The three of them managed the next blow without the door giving much ground.

Halcreek's blood dripped down onto the two men below him.

It could have been a few more seconds or even a lifetime of struggle before the outside went still. The ship was quiet, like the sirens had shifted into fog themselves in an instant. Still, the men held firm, scared that at any moment the door would come flying open and it would be the end of all of them.

Halcreek was the first to break his stance. He fell to his knees, sweating and pale. It was only clear now how much blood was on the floor around him—even the door was smeared heavily with his blood.

"Halcreek." Edmond tried to help him up. "Let's get you patched up."

Willas finally came away from the door to help them.

"I need stitches," he grunted. "I need to seal these cuts."

"I don't know how to do that." Willas showed his hands, looking confused.

"I can do it," Halcreek told them. "Just… just help me." It wouldn't be the first time he would have sewn himself up. Willas and Edmond were at his side with his equipment.

Halcreek's hands shook as he tried to thread his needle. The blood on his fingers was too slick and he ended up dropping the needle more than once. He let out a string of curses.

"Let me try." Edmond couldn't watch the man struggle anymore. He took the needle and placed the thread through it after a few missed attempts. Tensions were high. He couldn't exactly focus himself.

"My love?"

Edmond froze in place when he heard it. He looked at the other men. They were busy with one of Halcreek's arms.

"Come to me, my darling." Maryann sounded so sad. "Why do you hide from me?"

Edmond turned to look at the door. His love. His heart. His other half. She would listen to him. He could talk to her. They could find reason together.

Edmond drifted, making his way to the door now.

"Captain!" Willas called out. "Where are you going?"

Edmond couldn't tell him. He would only call him crazy, but Edmond knew better. He knew his Maryann. He knew she would listen. He only had to get to her.

He knew he could help his men.

He rushed to move the table and work his way out of the room, all while Halcreek and Willas called out for him to come back.

He followed the soft sound of her song until it led him above deck.

He found her looking flawlessly beautiful, her white gown flowing like mist around her. She looked angelic to his worn and lonely eyes.

"Maryann…" he gasped, taking in the lovely sight of this walking dream before him. His heart ached ever so slightly when he remembered he missed the birth. Where was his baby?

Suddenly, a vile, bulbous-eyed beast sprang at him from the fog. Edmond's back hit the deck hard, and the creature was on top of him. Gnashing teeth snapped away in front of his face while he shoved at the creature's chest to push them away.

Maryann let out a scream, her features twisting into a monstrous demon.

The scream was enough to startle the creature on top of Edmond.

Maryann's face softened again, her face smooth and gentle. She spoke in a language Edmond didn't understand. He watched her speak, then turned back to the creature that had been attacking him.

Instead of a siren on top of him, he now saw a beautiful woman.

This sight was just as troubling. He quickly crawled backwards to get away from her. She stood, but did not pursue him. The further he got from her, the more she disappeared into the fog.

He came closer to Maryann, stopping at her bare feet. He looked up to his wife with a sense of longing and fear.

She rested her pale hand on his shoulder. "Why do you hide them from me? After they took our daughter? I want them, my love."

She wanted the researchers. He knew it. He knew she wanted them. He wanted them off his ship just as badly now.

"Take them," Edmond found himself saying before he could even really process the full thought. Was he a murderer if he simply handed them over? In this moment, he no longer cared. "I can show you where they're hiding."

Chapter 22

Edmond lifted his hand, pointing down the hall towards the door with the hole fired through it. "There," he told her. "But they have a gun. Or guns. I'm not certain."

"We will handle them, my darling." She sounded determined and sure enough. Then, her tone changed to one of curiosity. "They've been locked away?" she asked, her voice a soft echo.

"Yes," he said. "We have no way of getting them out. We tried..."

"Then… who did it?" She eyed Edmond now, her tone suddenly sounding darker. "Who killed one of my daughters?"

Edmond's stomach turned. He had almost forgotten about the grisly scene he had witnessed not too long ago.

He felt his throat tighten up. He dared not even speak Willas's name. He couldn't throw his own man overboard like that. "I didn't…" He hesitated on what to tell her. "I didn't see who." Would she be able to distinguish lies from truth? Was it even worth trying?

"Don't lie to me, my love." She brushed her hand lovingly across his tired face.

He melted under her gentle touch. "Willas." He found the word leaving his lips before he could even realize he had made a sound. It had to be witchery. Or love. Or some combination of the two.

He couldn't fight it. His wife held his heart in her hands. This was somehow her, even if it wasn't.

"I want him." She made that point clear enough. She sounded so scornful, so pained by her loss. As if Edmond wasn't hurt enough by all he had lost.

"But..." Edmond felt his heart twist. "But you've taken so many already. Is that not enough? Have we not paid with enough blood, my dear?"

"That was for taking my daughter. I would have taken you all in order to get her back," she said. "You live now because you returned her to me unharmed."

"Willas helped. We all returned her to you!" Edmond pointed out. "He was defending himself! She attacked Halcreek! Your daughter—"

"I want him," she repeated. Her tone deepened to indicate there was no further debate.

"Captain?" Willas came around the corner now. He was on edge. His hands were covered in blood. Seeing her didn't soothe him at all.

"Willas," Edmond gasped softly, seeing the blood. "Halcreek?"

Willas hesitated to speak in front of the siren, eyeing it cautiously. "He is not good. He's losing a lot of blood. We need..." He lowered his voice as if it would make much of a difference before continuing, "We need help."

Edmond turned to the siren—his wife—and gave her a pleading look. "I have to help him."

She nodded. Then, she began her way to the door where the researchers hid. The men were screaming from inside, likely watching from the hole in the door.

Edmond didn't have time to concern himself with anything the sirens were doing. He needed to help Halcreek. He rushed after Willas back to the sick bay. Edmond had to step over the dead siren, his boot smearing some of the inky blood. They would likely want her returned to the sea. He would have to worry about that later.

"Halcreek?" Edmond touched his shoulder. The man was hunched over, covered thickly in his own blood. His head was dipped low.

"You still with us, buddy?" Willas brought over a bowl of water. He used a damp rag to wipe his face first, then to wipe away at his arm that they hadn't been able to stitch. Halcreek was just too weak. He grumbled, but it was barely audible.

"Hey." Edmond tried to help Halcreek lift his head gently.

He was so pale.

So very pale.

He couldn't lift his head on his own anymore. He was hunched over on himself, cradling his arms to his chest as if he were trying desperately to hold himself together. His mumbles went quiet. His eyes slowly closed.

"We can fix this." Willas tried to soothe the situation. "Halcreek, you're going to be okay!" Despite his words, it was clear that he couldn't help. Edmond couldn't help. No one could help. It was too late for Halcreek.

Soon, his chest no longer moved up and down. His arms loosened and slipped open to fall to his sides.

"Halcreek!" Edmond screamed, grabbing the man's shoulders. "Halcreek! Answer me!" He shook him lightly. Halcreek stayed limp and hunched, pale and cold. The blood loss was too much.

He was gone.

"No!" Willas cried out. His eyes were damp and glossy. He put his head to Halcreek's chest to listen for a heartbeat, only to come away with blood staining his blonde hair and a frown on his face. "No." He shook his head. "They killed him."

"This trip was a curse," Edmond said under his breath. Maryann had been right. He should have listened to her. He never should have left the harbor. He could be home now with his baby if it weren't for those damned researchers.

"What do we do?" Willas looked to his captain; a tear had escaped his eye to trail down his bloodstained face. Edmond could have cried, too, if he weren't so numb. There was nothing he could do about Halcreek, that much he knew, but they could still do something for themselves. The solution was easy enough. It was already out of their hands.

"We let them have them. They take them. Then we can go home." Edmond leaned against the wall, slid down it, then sat with his knees up in front of his chest.

Never in all his years did he think he would lose Halcreek. He was a good man. Ex-military. He had helped so many people, and now, he sat dead in the room he had helped so many others in. There was nothing Edmond could do to bring him back.

Edmond closed his eyes. He felt numb. He sat there only to listen to his breath and nothing more. He wanted peace—that was all he wanted.

If he could vanish from here and be in Maryann's arms again, he would be in heaven. He would be back at home. All his men would be safe and smiling in the sunlight. His ship wouldn't be busted and tattered.

None of this could be real. It must not have been happening. This must have all been a horrible nightmare that he hadn't woken up from yet. Why couldn't he just wake up?

Loud shots could be heard ringing out across the ship.

The researchers were fighting back.

Screeches were heard right after. Were the sirens injured, or was that a battle cry?

Edmond didn't know. He didn't care. It wasn't his battle. He and Willas stayed by Halcreek throughout the fight.

Crashes and splintered wood...

Shattered glass and more gunshots...

The researchers screamed what could be assumed to be curses as they fought for their lives.

The fight didn't last too long.

Finally, all was silent.

It was over.

Chapter 23

Edmond stepped into the room the researchers had once taken sanctuary in. The entire room was a disaster. There was certainly a struggle. Blood dripped down the wall under the jagged, broken window. It wasn't a very clean exit for the men. Edmond could only imagine what was happening to them now. It was clear enough that the sirens would show them no amount of mercy.

Willas approached the window, looking out cautiously. "The fog isn't lifting," he observed. "I thought they would let us go after they took them."

"Hm," Edmond grumbled. He couldn't bring himself to tell Willas what the mother siren had told him. Should he? They had been sailing this whole time and never outran the sirens. Was it truly impossible? Were they doomed to sail forever into nothingness until the sirens got Willas?

Edmond felt a knot in his throat. He couldn't even find the words to express how deeply and badly he wanted to go home. He was starving away to nothing. His clothes hung on his frame loosely. He was cold. Not just your average cold. He was cold to the bone; he could feel the frost within his very soul. He desperately wanted to feel the sun on his face again.

He looked at Willas. The man looked just as awful as Edmond felt. Dark circles were under his eyes, his face was more slender, his hair was a mess of dirt and blood, and his once bright blue eyes looked dull and lifeless. He looked to be a shell of the man he once was. Maybe…

Maybe he would be better off not remembering this trip?

Maybe it would be a blessing if he were taken by the sirens?

Edmond felt horrible for even thinking about it. Guilt ate away at his guts along with the hunger pangs.

Willas was his best friend, his first mate, his right hand, and all he had left.

So why did he envision throwing the man overboard?

"Can't you talk to them?" Willas asked. "You've reasoned with them before. Remind them of our exchange. Tell them to let us go." Poor Willas. He looked so hopeful, grasping onto any idea that they could eventually return home.

How many days and nights had they been sailing already? It all blended together. Edmond couldn't even tell what day it was anymore. His mind was strained and exhausted. Was it day or night even now?

"They won't let us go," Edmond sighed. "Not yet."

"What more could they possibly want from us?" Willas threw up his hands. "They've taken everything! There's no one left but us!"

"Hm." Edmond eyed the man again. He was so tired and weak—would he even be able to overpower Willas anymore?

"What?" Willas asked, giving his captain a strange look. After a moment, his eyes widened. "They want…" He paused in disbelief. "They want you, don't they? That creature that mimics Maryann. Or the one that takes the likeness of Emmalien." He shook his head. "They can't have you."

Willas approached Edmond and grabbed his shoulders. "I won't let them have you! Don't you worry! We'll get out of this, Captain! It's just us."

Willas embraced Edmond in a tight hug. Edmond felt even more guilty and sour within the embrace. He didn't deserve it. Willas was ready to fight for him while Edmond was struggling with the idea of sacrificing his best friend.

"It's just us," Edmond repeated, feeling cold and distant. He saw Maryann's face in his mind. He saw her holding a tiny bundle, waving at him in the harbor, the sun reflecting off her auburn hair. He saw her bright smile, excited to see her husband return. He was so close.

So very close.

All he had to do was…

He kicked himself again. How could he honestly be struggling with this? Willas would never consider such a thing, but Willas had no wife or children. He couldn't understand how desperate Edmond was.

While Edmond was desperate to return to his family, he realized that Willas fought so hard because Edmond… their men… Emmalien's Constellation was his family. They were all he had. That realization twisted the knife in Edmond's heart further. How could he betray Willas now? How could he betray him *ever*?

"What do we do?" Willas asked, still looking to his captain for guidance and answers. As if Edmond needed even more reason to feel guilty.

"There is nothing we can do," Edmond grumbled. "Nothing. We are powerless." He felt utterly defeated. "They will take what they want. All we can do is wait."

"We will starve," Willas pointed out, "until we can find land. Or another ship. They won't let us fish."

"No, they won't," Edmond agreed. He was having an impossible time finding any glimmer of hope.

"We could…" Willas shrugged, hesitating to finish. "We have one of their bodies…"

"We aren't eating it." Edmond shot that down while Willas was still in the middle of delivering the sentence. "We will return her body to the sea. We have to."

"I don't get it." Willas looked so confused. "If they wanted us dead, why haven't they just killed us yet? It is as if they are toying with us. Like they want to drive us insane first. To starve us. Dwindle us down. Build up our fear."

"You sound like you are the one researching them," Edmond said with a very weak chuckle.

"I just don't understand them," Willas scoffed. "Why do they want you?"

Edmond froze. He didn't want to answer. To answer would mean either confessing the truth or building up some lie to throw him off. Willas deserved better than that, but apparently not enough to know the truth. Besides, wouldn't that just lead him to have unnecessary fear? He did have one thing correct. If the sirens wanted someone dead, they would be dead. They would come for Willas eventually. Edmond just had to withhold the warning. Was he doing his best friend a favor or a disservice?

Was it truly betrayal if there was nothing Edmond could do to stop it, anyway?

"I don't know." Edmond shook his head. "I don't know what they want."

"Unless they lied," Willas offered. "This whole thing was a game to them. They meant for us to go crazy before they finally climbed up here and killed us. Devourers of men—that's all they are. Perhaps we should have seen this coming."

Was that the case? Would they even let Edmond go after they took Willas? His Maryann wouldn't lie to him, would she? He shook his head, trying to clear his mind of the thoughts. It wasn't even Maryann. Why did he keep getting them confused? It was a slimy creature of the ocean, not his warm and loving wife.

"Looks like it is useless to barricade ourselves," Willas continued. "At this point, we simply aren't strong enough to stop them."

"Let's give their dead back to them," Edmond told him. "Maybe that is what they are waiting for." He pictured it in his mind. When they tossed her body over the side of the ship, Willas would "accidentally" slip and tumble into the ocean. He was so trusting, he would never suspect that Edmond would…

Edmond rubbed his face hard. Why did these scenes keep coming to him? Was he in such a hurry to kill Willas? The man was practically his brother. All he had to do was wait—but how long?

"Let's do that," Willas agreed.

Together, they worked the slimy, wet, cold, and limp body of the siren onto a sheet. Touching it did nothing to their senses. Their powers stopped completely at death. So, maybe they weren't devils or demons—just flesh and bone, like the researchers said.

They carried her above deck, careful to balance her between them. Edmond counted every step they took. Each step brought them closer to the edge. Each step potentially brought him closer to Maryann. Did he have what it would take?

"Right here should be good." Willas came first to the edge.

"Yes." Edmond came to stand against the edge as well. He looked down at the choppy waters. Were the sirens just below, swimming alongside them? Was Emmalien there? Would they know that they were returning this siren to them? Is that what they wanted, or was it seen as disrespectful to toss her off?

It was hardly the time to be thinking it over now.

They were right here.

It was time to toss her over.

Together, they lifted the beast up, realizing it was far heavier than they originally anticipated. The sheet came to rest on the edge now.

"On three," Edmond instructed. He looked to Willas and the man nodded in return. On the count of three, Willas would shove his own body forward with enough momentum. It would only take a small push…

"One." Edmond steeled himself.

"Two." He eyed Willas's back.

"Three." He shoved forward with all his might.

Chapter 24

Edmond shoved the carcass with all his might. It plummeted into the sea with a mighty splash.

"Hopefully, that will show them we mean well," Willas sighed, leaning his elbows along the edge of the ship.

Edmond tried his best to swallow the knife in his throat. He looked over at Willas. As much as he wanted to go home, he couldn't betray Willas. He simply couldn't. He felt shame like he never had before. To stoop so low... Edmond wasn't a murderer.

He would have to reason with the mother siren. Surely, he could. Halcreek was dead—that had to be enough payment. An eye for an eye. It should be done. He would just need to do his best to express this to her.

It was a wonder they hadn't returned yet. Perhaps they were still busy with the researchers. *May God have mercy on them.*

Edmond sighed and clasped Willas's shoulder. In this moment, Edmond couldn't be more thankful that the spell of madness hadn't won him over. He had never felt such a pull to do such wrong. He hoped that spell had passed for good.

Willas was willing to fight for Edmond. Edmond should be willing to fight for Willas. Hopefully, it wouldn't come to that. Either way, he was certain that in this moment, he was willing to put his life on the line for his best friend.

"Does the fog look any lighter?" Willas half joked. This man wanted to go home just as badly as Edmond did, but it was obvious the fog wasn't going anywhere.

"They aren't finished with us yet." Edmond looked down at the water. If only Emmalien were here. She would listen. She would vouch for him. He knew he had made a real connection with her.

"To the death?" Willas looked over the waves, too. It was a hollow sentence. A cold one. One that neither of them thought they would have to embrace. But if Edmond couldn't wake up, then this was their reality.

He thought of his wife again. He only hoped she wouldn't be too heartbroken. He hoped she would be safe and well. His salary should have been enough to sustain her and the baby for a short time. After that, he wasn't sure. What would become of his family? Maybe they would move in with her mother. Her mother would curse Edmond for being too money hungry to stay with his wife. Maybe she would be right, but he knew he was doing this for their betterment. Hopefully, Maryann knew that in her heart, too.

Edmond leaned on the side of the ship as well. The sky was getting darker. Night was upon them. Would that mean the next attack?

He was so exhausted. He felt like he hadn't slept in days. Had it been days? Weeks? Months? His loss of time unsettled him. He went below deck to update his logbook.

I have lost track of the days. They all blend together.

Halcreek is dead. He was killed by a siren. Willas... We almost lost him to the sea. By my own hand. I almost sacrificed my best friend to them in order to return home.

They aren't brainless beasts. I can reason with Maryann. I can reason with Emmalien. If they should return.

If they will not listen to reason...

To my wife. My Maryann. I love you more than life itself. Not a day goes by that I don't long for you. I am sorry I didn't make it home for the birth. I know you are caring for our little one well. Our little Edmond junior… or our Edna. Sorry we never got to discuss names.

I hope you can find it in your heart to forgive me. To still love me. Please remember that I will always be with you. I will watch over you. Find a reason to smile every day.

Yours always, Edmond Conway

He leaned back in his chair and had to rub his eyes to smear away the tears that burned his skin. He longed to touch her face just one last time, to hold her, to kiss her lips once more. He tried his best now to remember how she tasted, to remember how tender her lips were. If he had known his kiss goodbye was the last kiss he would ever get, he would have saved it deeper into his memory. How did she taste? He could almost feel her warm lips on his.

"Should we try to barricade anyway?" Willas tried to ease the tension of the moment.

"I need to talk to them," Edmond said. He couldn't say much more on the topic without giving anything away.

"Let's hope they will hear you out." Willas took a moment to look out the window, only to discover the fog leaking back into the ship.

They were coming.

"Find somewhere to hide," Edmond told him, standing quickly. "I'll talk to them."

"I'm not leaving you alone with them!" Willas protested. "They will take you!"

"That's a risk I'm willing to take," Edmond assured him. "Now, hide. Before it's too late."

"I'm with you," Willas demanded. "Always."

"And if they take you?" Edmond felt frustrated arguing when they didn't have time.

"If I die so you can go home? So be it," he declared.

Edmond froze again. Did he mean that? How could he say such a thing? Edmond thought back to the split second decision he made to spare Willas, wondering if Willas would have willingly gone if only he had asked him to.

He didn't want to think about it. He had lost all of his men already. Willas was all that was left. He couldn't lose him, too.

"Please hide," Edmond tried one last time, already knowing the outcome.

"To the death," Willas reiterated and headed out of the room.

Edmond followed him to the deck. Why hide anyway? They would be found out eventually. It was best to just get this encounter over with.

They had to stay close together so neither were swallowed by the fog.

The familiar sound of wet clawed hands slapping the wood sounded.

They were climbing the ship, perhaps for the last time.

All was silent now.

The men stood back-to-back, looking in every direction just to see nothing.

"Maryann!" Edmond called out. "Please! Listen to me!"

Unfortunately, Edmond calling out only seemed to put a clear target on their exact location. In half a second, both of them were tackled. Edmond slammed onto his stomach. Willas landed on his back, pushing him further onto the deck.

"Wait!" he gasped out.

A screech was heard. Was it Maryann giving her commands?

They were tackled again, both of them rolling across the deck. Willas had a slash across his arm that was deep.

"Captain." Willas clutched his arm tightly. "It was an honor to sail with you!"

"Don't start!" Edmond cried desperately.

Just then, Willas was yanked backwards into the fog, the mist swirling in a hurry to fill the now empty space.

"No!" Edmond jumped up and ran blindly into the fog as fast as he could. Had he failed so suddenly?

He saw it. The creature was dragging Willas to the edge. It had a hand over the edge now. Willas struggled to free himself.

"Stop!" Edmond begged, closing the gap.

The creature used its free hand to slash deeply across Edmond's chest. As it turned to jump over the edge at last, it whipped its serpentine tail back and slammed it into Edmond.

As Edmond flew back, he heard Willas screaming as he descended into a splash.

Edmond hit the deck so hard that it knocked the wind clean out of his lungs.

He gasped to take a breath, only to make pained, empty noises instead. He simply couldn't catch his breath. The hit was harder than he realized. A small puddle of blood was forming under his scalp. His vision blurred.

And then everything went dark.

Chapter 25

The wood swayed under him. He could feel the rocking of the ship. It was soothing, to a degree. Yet, he felt dizzy and nauseous. The world seemed to swirl just beyond his closed eyelids.

He heard the water lapping against the side of his ship. How long had he been out? How far had the wind taken him?

Slowly, very slowly, he peeled open his eyes.

A blurry mess was coming together in his vision. Beyond the wood, beyond the sails, there was a sky. A clear sky with more stars than he could possibly count.

He blinked a few times to try and focus his vision further. Dead ahead of him, he saw it just like he remembered it: Emmalien's Constellation. Pointing, just like any other constellation, how to make it home.

He breathed. He allowed his chest to rise and fall ever so slowly, taking in the stunning view. He hadn't seen the stars in so long. They were like little lighthouses in a vast blue ocean, just like Emmalien used to say.

Edmond's lips curled into a half smile. Was he truly free? Was this too good to be true? He sighed again as his vision blurred once more. His eyes felt heavy and forced. They burned. He was exhausted. He couldn't keep his eyes open much longer.

He gently let his eyelids come to rest again. Listening to the waves and wind in his sails, everything soon went dark once more.

He dreamed. Dreamed of Halcreek. Dreamed of Willas. Dreamed of DeVitas and Piles and all his crew.

They shared a hearty meal together. All around the table, his men laughed and chatted. A few sang a shanty together with their mugs held high in the air.

Edmond couldn't help but laugh. This is how things should have been. Maybe he was finally awake, and the rest had been a dream.

A two-legged Piles and Davis engaged in an arm wrestle. It was almost comical how much they both struggled against each other, both desperate to be the champion.

Edmond found Willas singing along with Shaw and Keller.

He wanted to hug the man but found that he wasn't in control. All he could do was watch all of his men be happy and at peace. Despite that, he could physically feel the warmth of the moment.

While the scene started to fade out, the warmth still lingered with him, along with the distant chatter.

This time, when Edmond came to, his face and chest still felt warm. His hand twitched, hitting the damp wood beneath him.

A light stained his eyelids yellow and orange.

Wincing, he did his best to open his eyes once more. This time, he opened them to see the sun shining brighter than he could swear it ever had before.

"Captain?" someone called out to him amongst the chatter.

Edmond was too weak to lift or even turn his head.

"That's Captain Edmond," someone else said.

Voices were heard around him, all muttering and blending together while Edmond did his best to keep his heavy eyes open. He felt stiff and frozen. Only the warmth of the sun told him he was still alive. The fact that he could feel the heat.

"Is that him?" another voice asked above the others.

Edmond didn't recognize the voices. He wasn't even certain if they were real or if they were somehow part of his imagination.

"Edmond!" a woman called out to him. "My darling!" Now, this was a voice Edmond would recognize even at the cusp of the void. His Maryann.

He felt a warm hand on his cool cheek. "My love! Edmond, please! Talk to me!"

His eyes opened again. This time, he could see her. Maryann. More beautiful than ever. Her hair blew in the wind over her face. Her eyebrows were creased in worry. "Edmond!" She patted his face lightly and took hold of his hand in hers. She clutched his hand to her heart like her life depended on it.

"Maryann…" Edmond said softly. He fought with all his remaining strength to lift his free hand to her face. He brushed her hair away from her eyes. He had to see her fully. Her bright eyes looked just as concerned as her half smile.

"My love." He stroked her face. "I… came back to you."

"Yes!" Her thumb gently rubbed his cheek. "You came back! You came back to me!"

"I said I would." He tried his best to smile now, still feeling drowsy.

"Never leave me again!" she demanded sharply.

He couldn't even blame her. He would have laughed if he had the strength and if it didn't hurt so much. "Never," was all he could say in reassurance.

"You'll never leave me again?"

"Never," he repeated.

That was all she needed to hear. Her smile made Edmond's heart melt. She grabbed his face with both her hands and kissed him firmly on the lips.

Edmond's heart could have exploded. It was everything he had wished for and more.

Peace.
He finally felt peace.
He was home at last.

Chapter 26

Maryann stood on the dock, watching the ship come into the harbor. It looked… wrong. So damaged and tattered. Had there been a storm? The anticipation ate away at her. She hadn't seen her husband in months. He was supposed to be home three months ago.

She paced anxiously, back and forth, holding her little baby bundled in a blanket within her arms. The baby cooed, not knowing the weight of the heavy situation. This moment would determine how the rest of their lives would go.

She waited with high nerves as the ship was docked and a handful of men made their way off the ship. It was more than she could bear. She had to find Edmond. She needed to throw her arms around him. He needed to hold their baby for the first time.

She rushed forward to meet the men who had stepped off the ship. She searched all of their faces. All of them grim, none of them Edmond.

"The ship is in ruins," the man at the lead told her. "Shattered windows. Busted doors. Splintered wood. Every room was ransacked."

"A…" She sounded confused, yet hopeful. "A storm?"

"More like an attack," the man said matter-of-factly. "Lots of blood. Everywhere. There was definitely a struggle in many places."

"Pirates?" she gasped, feeling horror wash over her.

"Likely," he said sadly. "We found Emmalien's Constellation drifting a few miles out. It was likely they were attacked on their way back home."

"Edmond?" She had dreaded to ask this more than anything, dreading the answer she already knew was coming. "Where is my husband?"

The man let out a long sigh before reluctantly answering, "There wasn't a single soul on board the ship. Who knows what happened to the crew? From the blood, I wouldn't guess it is anything good."

"He's… gone?" She didn't believe it. She would have to search the entire ship top to bottom in order to be sure herself.

"Indeed," the man said sadly. "Found this, though." He presented Maryann with the logbook. "I read it. But…" He trailed off.

"But what?" she asked, snatching the book with her free hand as if it was the key to finding her husband.

"I think it's possible that Captain Conway was sick from the sea. That he lost his mind to some sort of madness. It could have even been…" He paused before finishing, "mutiny."

"Never." She scolded the man to have even muttered such a thing, "Those men have sailed with my husband for years. They trusted him with their lives! They never would have done something like that."

The man interrupted her to tap the logbook. "Mrs. Conway, it's all in there. You can witness his mind deteriorating. He wrote a letter to you, though. On the last page."

Maryann held her baby out for the man to help her free up her hands. He took hold of the infant while she flipped through the pages. When she came to rest on the last page, her eyes scanned over the words, slowly filling with tears as she read.

She dropped to her knees as sobs took hold of her and rocked her shoulders. She was so confused, lost, and hurt at the same time. How could this be? Sirens? Had her husband truly lost his sanity somewhere out on the sea? He spoke of their Emmalien. Surely, something wasn't right, but what?

She knew.

She knew her husband wasn't a madman. She knew his words must have been the literal truth. Surely, these things happened. All his men were taken, including him.

She cast her eyes to the horizon, watching the rippling waves eat away at the skyline.

Her husband was out there somewhere.

That night, she lowered her little Edna into her crib and clutched the logbook to her chest. This was the final piece of Edmond she had left. Holding it felt almost like holding his hand.

She had read the book through from start to finish multiple times. Each time, she became angrier. Angry at the sea, angry at Edmond for not listening, and angry at the things she knew had taken him.

She sat in her rocking chair by the window. It was the same window she always sat by when she waited for her husband to return to her. A single candle was lit next to her—a sort of light she left on, as if it would guide him back home to her.

She looked out the window and glared at the broken ship that had taken her husband away. She knew it was cursed now.

She knew getting it repaired was an outrageous idea.

Yet, she was determined.

Soon, all would be in order.

She knew Edmond was out there somewhere.

She could feel it.

She was going to find her husband.

www.ingramcontent.com/pod-product-compliance
Lightning Source LLC
LaVergne TN
LVHW022056200926
848481LV00003B/944
9798992775433